FLUIDS FOR POWER SYSTEMS

R.H. WARRING

TRADE AND TECHNICAL PRESS LTD.,
MORDEN, SURREY, ENGLAND

85461 040 5

PRINTED AND PUBLISHED BY TRADE AND TECHNICAL PRESS LTD.

CONTENTS

1. PRESSURISED FLUIDS

A fluid is by definition a substance which can flow, the particles comprising the substance being readily displaced. A liquid is merely a category of fluid which, at a given temperature, occupies a definite volume and is largely unaffected by changes in pressure. Whilst 'hydraulics' is, again by definition 'pertaining to fluids', the fluids employed are invariably liquids; but equally invariably referred to as fluids. The term fluid will, therefore, be employed throughout as descriptive of (hydraulic) liquids.

The perfect or ideal fluid is incompressible and also non-viscous (ie it experiences no shearing stresses or internal friction). Classic hydrodynamics, or analysis of the behaviour of fluids in motion, is based on the behaviour of an ideal fluid. Any real fluid is both compressible and has finite viscosity, so that its behaviour is somewhat modified compared with that of an ideal fluid. This presents certain problems in applying ideal flow theory to practical design problems, although suitable solutions may differ in detail only – ie specifically correlation between theory and practice may often be achieved by the insertion of an empirical coefficient, or a modified exponential value, in an ideal fluid formula. In other cases only empirical data may be valid for specific applications; but over-emphasis of the 'practical approach' can be just as misleading at times, particularly if such data are extrapolated.

Flow Velocity

Flow velocity is determined on the basis that a fluid mass is a continuum. Thus where a pump delivers a specific quantity of fluid (Q) into a pipe the flow velocity through the pipe is determined by simple geometry, viz

$$V_1 = \frac{4Q}{\pi d_1^2}$$

At a change of pipe section, after an initial disturbance, flow will become established at another velocity, namely

$$V_2 = \frac{4Q}{\pi d_2^2}$$

If the initial flow is generated by a piston then the delivery will be related to the piston area and speed, viz

$$Q = \frac{\pi D^2 V}{4}$$

where D = piston diameter

V = piston speed

Flow velocities at different pipe sections following would then be

$$V_1 = \frac{D^2 V}{d_1^2}$$

$$V_2 = \frac{D^2 V}{d_2^2}$$

In other words, calculation is always based on the assumption that the flow completely fills the pipe when flow velocities (for a given delivery) are in inverse proportion to the square of the pipe bore. It should be noted that there are nominal or mean flow velocities. In fact, a velocity gradient will exist across the bore, determined by the flow condition (and also flow velocity in the case of turbulent flow). Localised high velocities may also occur, eg at bends, and stagnation areas at abrupt changes in section.

TABLE 1.1A.
FLOW RATES THROUGH SWG WALL TUBING

Wall Thickness Swg	Outside Diameter — inches										
	1/4	5/16	3/8	7/15	1/2	9/16	5/8	11/16	3/4	7/8	1
24	.3996	.6797	1.0332	1.4592	1.9608	7.5332	3.1812	3.9024	4.6980	—	—
22	.3552	.6204	.9600	1.3716	1.8588	2.4180	3.0416	3.7584	4.5396	6.3216	—
20	.2988	.5448	.8640	1.2588	1.7280	2.2680	2.8824	3.5700	4.3332	6.0780	8.1168
18	.2232	.4416	.7344	1.0992	1.5384	2.0508	2.6376	3.2976	4.0408	5.7192	7.7016
16	.1404	.3216	.5760	.9024	1.3056	1.7796	2.3280	2.9508	3.6468	5.2596	7.1664

TABLE 1.1B.
FLOW RATES THROUGH INCH BORE SIZE PIPING

Bore Size	1/8"	3/16"	1/4"	5/16"	3/8"	7/16"	1/2"	9/16"	5/8"	11/16"	3/4"	13/16"	7/8"
Flow Rate Factor*	.147	.331	.589	.920	1.33	1.80	2.36	2.74	3.70	4.41	5.30	6.72	7.21

1 5/16"	1"	1 1/16"	1 1/8"	1 3/16"	1 1/4"	1 5/16"	1 3/8"	1 1/2"
8.28	9.43	10.64	11.93	13.29	14.52	16.24	17.82	21.20

NOTE: flow rate is given in cubic inches per second for a flow velocity of 1 ft/sec. Thus:—

(i) To find the flow rate in cu.in/sec, multiply by actual flow velocity in feet per second.

(ii) To find the flow rate in gallons per minute, multiply table figure by 4.62 x flow velocity in feet per second.

(iii) To find the flow rate in cc per second, multiply by 16.4 x flow velocity in feet per second or 5 x flow velocity in metres per second.

(iv) To find the flow rate in litres per minute, multiply by 21 x flow velocity in feet per second or 6.4 x flow velocity in metres per second.

TABLE 1.C
FLOW RATES THROUGH DECIMAL SIZE PIPES

Bore	Flow Rate Factor	Bore	Flow Rate Factor	Bore	Flow Rate Factor	Bore	Flow Rate Factor	Bore	Flow Rate Factor
.1	.094	2.1	41.56	4.1	158.4	6.1	350.7	8.1	618.4
.2	.377	2.2	45.61	4.2	163.3	6.2	362.3	8.2	633.7
.3	.847	2.3	49.86	4.3	174.3	6.3	374.0	8.3	649.2
.4	1.51	2.4	54.29	4.4	182.5	6.4	386.0	8.4	665.0
.5	2.12	2.5	58.91	4.5	190.0	6.5	398.2	8.5	681.0
.6	3.39	2.6	63.71	4.6	199.4	6.6	410.5	8.6	697.0
.7	4.61	2.7	68.71	4.7	208.2	6.7	425.1	8.7	713.3
.8	6.03	2.8	73.88	4.8	217.2	6.8	435.8	8.8	729.8
.9	7.63	2.9	79.26	4.9	226.3	6.9	448.7	8.9	746.5
1.0	9.43	3.0	84.83	5.0	235.6	7.0	461.8	9.0	763.3
1.1	11.40	3.1	90.58	5.1	245.1	7.1	475.1	9.1	780.4
1.2	13.57	3.2	96.50	5.2	254.8	7.2	488.6	9.2	797.6
1.3	15.93	3.3	102.6	5.3	264.7	7.3	502.2	9.3	815.1
1.4	18.47	3.4	108.95	5.4	274.8	7.4	516.1	9.4	832.8
1.5	21.20	3.5	115.5	5.5	285.1	7.5	530.2	9.5	850.6
1.6	24.13	3.6	122.1	5.6	295.5	7.6	544.4	9.6	868.6
1.7	27.34	3.7	129.0	5.7	306.2	7.7	558.8	9.7	886.7
1.8	30.54	3.8	136.1	5.8	317.0	7.8	573.4	9.8	905.1
1.9	34.02	3.9	143.3	5.9	328.1	7.9	588.2	9.9	923.7
2.0	37.70	4.0	150.8	6.0	339.3	8.0	603.2	10.0	942.5

NOTE: For bore size in inches, multiply flow rate factor by flow velocity in ft/sec for flow rate in cu.in/sec.
" " " " " 4.63 x flow velocity in ft/sec for flow rate in gpm.

for bore size in millimetres, multiply flow rate factor by 0.83 x flow velocity in metres/sec. for flow rate in cubic centimetres/second.

TABLE 1.D.
UNIVERSAL PIPE SIZING TABLE

Flow Rate	FLOW VELOCITY									
	1	2	3	4	5	6	7	8	9	10
1	1.0000	5000	.3333	.2500	.2000	.1667	.1429	.1250	.1111	.1000
2	2.000	1.000	.6667	.5000	.400	.3333	.2858	.2500	.2222	.200
3	3.000	1.500	1.000	.7500	.600	.5000	.4287	.3750	.3333	.300
4	4.000	2.000	1.333	1.000	.800	.6667	.5715	.5000	.4444	.400
5	5.000	2.500	1.667	1.250	1.000	.8333	.7143	.6250	.5555	.500
6	6.000	3.000	2.000	1.500	1.200	1.000	.8574	.7500	.6666	.600
7	7.000	3.500	2.333	1.750	1.400	1.167	1.000	.8570	.7777	.700
8	8.000	4.000	2.667	2.000	1.600	1.333	1.143	1.000	.8888	.800
9	9.000	4.500	3.000	3.250	1.500	1.500	1.286	1.1250	.9999	.900
10	10.000	5.000	3.333	2.500	2.000	1.667	1.429	1.250	1.111	1.000
11	11.000	5.500	3.667	2.750	2.200	1.833	1.572	1.375	1.222	1.100
12	12.000	6.000	4.000	3.000	2.400	2.000	1.715	1.500	1.333	1.200
13	13.000	6.500	4.333	3.250	2.600	2.167	1.858	1.625	1.444	1.300
14	14.000	7.000	4.667	3.500	2.800	2.333	2.000	1.750	1.555	1.400
15	15.000	7.500	5.000	3.750	3.000	2.500	2.143	1.875	1.666	1.500
16	16.000	8.000	5.333	4.000	3.200	2.667	2.286	2.000	1.777	1.600
17	17.000	8.500	5.667	4.250	3.400	2.833	2.429	2.125	1.888	1.700
18	18.000	9.000	6.000	4.500	3.600	3.000	2.572	2.250	1.999	1.800
19	19.000	9.500	6.333	4.750	3.800	3.167	2.715	2.375	2.111	1.900
20	20.000	10.000	6.667	5.000	4.000	3.333	2.858	2.500	2.222	2.000
21	21.000	10.500	7.000	5.250	4.200	3.500	3.000	2.625	2.333	2.100
22	22.000	11.000	7.333	5.500	4.400	3.667	3.143	2.750	2.444	2.200
23	23.000	11.500	7.667	5.750	4.600	3.833	3.286	2.875	2.555	2.300
24	24.000	12.000	8.000	6.000	4.800	4.000	3.429	3.000	2.666	2.400
25	25.000	12.500	8.333	6.250	5.000	4.167	3.572	3.125	2.777	2.500
26	26.000	13.000	8.667	6.500	5.200	4.333	3.715	3.250	2.888	2.600
27	27.000	13.500	9.000	6.750	5.400	4.500	3.858	3.375	2.999	2.700
28	28.000	14.000	9.333	7.000	5.600	4.667	4.000	3.500	3.111	2.800
29	29.000	14.500	9.667	7.250	5.800	4.833	4.143	3.625	3.222	2.900
30	30.000	15.000	10.000	7.500	6.000	5.000	4.286	3.750	3.333	3.000
31	31.000	15.500	10.333	7.750	6.200	5.167	4.429	3.875	3.444	3.100
32	32.000	16.000	10.667	8.000	6.400	5.333	4.572	4.000	3.555	3.200
33	33.000	16.500	11.000	8.250	6.600	5.500	4.715	4.125	3.666	3.300
34	34.000	17.000	11.333	8.500	6.800	5.667	4.858	4.250	3.777	3.400
35	35.000	17.500	11.667	8.750	7.000	5.833	5.000	4.375	3.888	3.500
36	36.000	18.000	12.000	9.000	7.200	6.000	5.143	4.500	3.999	3.600
37	37.000	18.500	12.333	9.250	7.400	6.167	5.286	4.625	4.111	3.700
38	38.000	19.000	12.667	9.500	7.600	6.333	5.429	4.750	4.222	3.800
39	39.000	19.500	13.000	9.750	7.800	6.500	5.572	4.875	4.333	3.900
40	40.000	20.000	13.333	10.000	8.000	6.667	5.715	5.000	4.444	4.000
41	41.000	20.500	13.667	10.250	8.200	8.833	5.858	5.125	4.555	4.100
42	42.000	21.000	14.000	10.500	8.400	7.000	6.000	5.250	4.666	4.200
43	43.000	21.500	14.333	10.750	8.600	7.167	6.143	5.375	4.777	4.300
44	44.000	22.000	14.667	11.000	8.800	7.333	6.286	5.500	4.888	4.400
45	45.000	22.500	15.000	11.250	9.000	7.500	6.429	5.625	4.999	4.500
46	46.000	23.000	15.333	11.500	9.200	7.667	6.572	5.750	5.111	4.600
47	47.000	23.500	15.667	11.750	9.400	7.833	6.715	5.875	5.222	4.700
48	48.000	24.000	16.000	12.000	9.600	8.000	6.858	6.000	5.333	4.800
49	49.000	24.500	16.333	12.250	9.800	8.167	7.000	6.125	5.444	4.900
50	50.000	25.000	16.666	12.500	10.000	8.333	7.143	6.250	5.555	5.000

This table gives the theoretical cross section of pipe required for a given rate and given flow velocity, in consistent units—viz.:—

cross sectional area given in:—	for flow rate in:—	and	flow velocity in:—
square inches	cu-in/sec.		inches/sec.
sq. centimetres	c.c./sec.		centimetres/sec.

To correct for practical flow velocity units apply the following to the tabular answers.

For cross section in sq. In. and flow rate in ft./second, divide by 12.

For cross section in sq. in. and flow rate in ft./minute, multiply by 5.

For cross section in sq. cm. and flow rate in metres/second, divide by 100.

Other correction factors are readily determined for working in any other set of units.

Fluids for power systems are concerned primarily with fluids under pressure. For most practical purposes the velocity of transmission of pressure through a fluid can be considered instantaneous. Its velocity (V_P) is given by the formula

$$V_P = \sqrt{\frac{Kg}{\rho}}$$

where K is the bulk modulus of the fluid
g = acceleration due to gravity
ρ = density of fluid
(in consistent units).

In English units the bulk modulus of water is of the order of 43.2×10^6 lb/ft², and that of a typical hydraulic oil around 40×10^6 lb/ft²; giving a velocity of transmission of pressure of approximately 4700 ft/sec (1 430 metres/sec) and 4840 ft/sec (1 470 metres/sec) respectively.

Compressibility

The bulk modulus of a fluid also defines its 'compressibility'. Specifically, compressibility (β) is the reciprocal of the bulk modulus (K). Its value is not a constant, however, for it tends to increase with temperature and decrease non-linearly with pressure – see Tables 1.2 and 1.3.

TABLE 1.2

COMPRESSIBILITY – TEMPERATURE RELATIONSHIP FOR A TYPICAL LIGHT MINERAL OIL. TABLE FIGURES GIVE RELATIVE VOLUME

Pressure		Temperature °C					
psi	kg/cm²	–40	–20	0	20	40	60
14.7	1	.952	.970	.988	1.004	1.020	1.033
1000	70	.949	.967	.983	.997	1.014	1.024
2000	141	.947	.964	.978	.992	1.007	1.018
3000	211	.945	.961	.974	.987	1.000	1.012
4000	281	.942	.958	.970	.983	.995	1.006
5000	352	.940	.955	.967	.978	.991	1.000
6000	422	.938	.952	.964	.974	.988	.996
7000	492	.936	.949	.961	.970	.985	.993
8000	562	.935	.947	.958	.967	.981	.990
9000	633	.933	.945	.955	.965	.978	.986
10000	700	.931	.942	.952	.963	.975	.983
11000	770	.930	.940	.950	.960	.972	.980
12000	840	.929	.938	.948	.957	.969	.977

TABLE 1.3

COMPRESSIBILITY – PRESSURE RELATIONSHIP (COMPRESSIBILITY EXPRESSED AS PERCENTAGE REDUCTION IN VOLUME)

FLUID	% Reduction in Volume at 1000 psi (70 kg/cm^2)	10000 psi (700 kg/cm^2)
Water	0.34	3.3
Water-in-oil Emulsions	0.35	3.5
Water-Glycol	0.26	2.6
Mineral oil (typical)	0.35	3.4
Phosphate Ester	0.25	2.5
Chlorinated Hydrocarbon	0.24	2.4

As a consequence, the bulk modulus can be defined in two ways, each of which can provide an isothermal or isentropic value. The instantaneous value of the bulk modulus at any pressure is called the *tangent modulus*, (Fig 1.1) given by

$$K = -V\frac{dP}{dV}$$

where V is the volume at pressure P

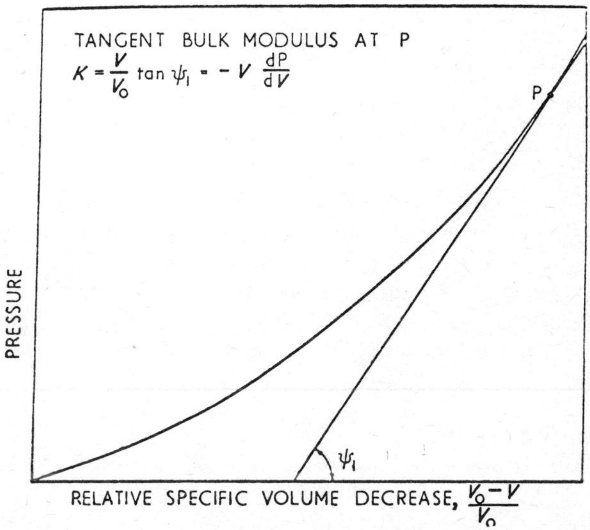

Fig. 1.1.

The mean value of the bulk modulus from atmospheric pressure to pressure P is called the *secant modulus,* given by

$$K = -\frac{V_0 P}{V_0 - V}$$

where V_0 is the volume at atmospheric pressure (See also Fig 1.2).

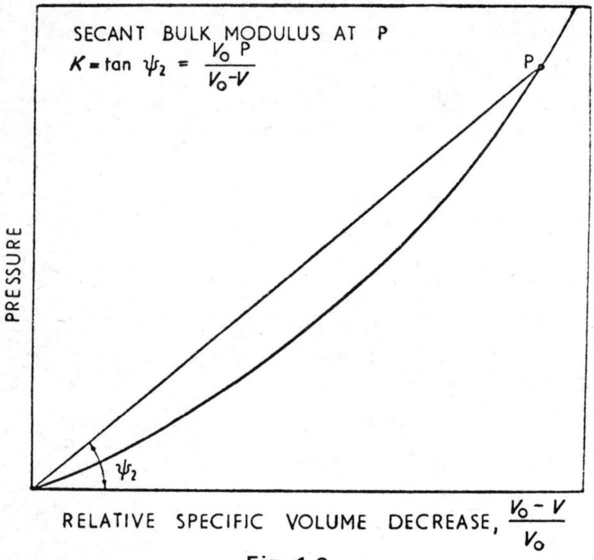

Fig. 1.2.

The values of the tangent and secant moduli will tend to coincide at lower pressures (ie as P approaches atmospheric pressure), and for pressures up to about 1000 psi the difference can usually be ignored. For general engineering calculations a 'typical' bulk modulus value may be quoted and used for pressures up to 10 000 psi. Logically this should be the secant modulus. For working at specific high pressures, however, the tangent modulus should be used, if known.

In many practical cases the distinction is largely academic since the apparent compressibility is larger than the actual fluid compressibility. This is due to the elasticity of the system itself yielding an effective increase in internal capacity. 'Typical' bulk modulus values may thus be derived with this in mind, giving higher values than true fluid compressibility figures. Such general figures are also usually quoted in terms of 'compressibility' rather than bulk modulus, defined as

$$\text{'compressibility' (per cent = reduction in volume)} \quad \frac{V_o - V}{V_o} \times 100$$

$$\text{or} = \frac{V_o - V_{1\,000}}{V_o} \times 100 \text{ per cent per } 1\,000 \text{ psi}$$

where $V_{1\,000}$ = volume at a pressure of 1 000 psi

Typical working figures are a reduction in initial volume of 0.5 per cent per 1 000 psi (0.00735 per cent per atmosphere) for typical hydraulic oils; and 0.4 per cent per 1 000 psi (0.0059 per cent per atmosphere) for water. Such figures are reasonably valid for pressures up to 10 000 psi (700 kg/cm^2) and over a temperature range of 50 to 200 degrees F (10 – 100 degrees C).

The compressibility of a normal hydraulic oil at 20°C(68°F) and 1 000 psi can also be estimated quite accurately from its kinematic viscosity at 22 degrees C (72 degrees F), using the empirical formula

$$\text{'compressibility'} = 3.5 - 0.2 \log \nu$$

where ν = viscosity in centistokes

and compressibility is given in inch^2/lb $\times 10^{-6}$ units

Alternatively, 'compressibility' may be expressed in terms of relative density, viz

$$\text{'compressibility'} = (i - \frac{\Delta_{1\,000}}{\Delta_o}) \times 100 \text{ per cent per } 1\,000 \text{ psi}$$

where Δ_o = fluid density at atmospheric pressure

$\Delta_{1\,000}$ = fluid density at 1 000 psi

As a direct result of compressibility, the density of any real fluid will increase with pressure. In very high pressure systems this may be more significant than the volumetric change. Typical data for a range of fluids are summarised in Tables 1.4 and 1.5.

The main problem associated with compressibility is that it can affect the accuracy of positioning or precise movement achieved in high pressure systems, and can also lead to flutter or instability under varying load conditions. It is normally only significant in high-pressure, high-duty systems.

TABLE 1.4

TYPICAL PRESSURE – DENSITY RELATIONSHIP (AT CONSTANT TEMPERATURE) DENSITY IN GRMS/CC OR EQUIVALENT TO SPECIFIC GRAVITY

FLUID	Pressure psi (kg/cm^2)					
	0 (1)	2 000 (140)	4 000 (280)	6 000 (420)	8 000 (560)	10 000 (700)
Mineral Oil	0.842	0.851	0.858	0.864	0.870	0.876
Water-Glycol	1.08	1.09	1.09	1.10	1.10	1.10
Castor-Methanol	0.943	0.952	0.959	0.966	0.971	0.977
Phosphate-Ester	1.067	1.076	1.083	1.089	1.094	1.099

TABLE 1.5

TYPICAL PRESSURE – RELATIVE DENSITY RELATIONSHIP FOR A MINERAL OIL FLUID AT 20°C

Pressure psi	1 000	2 000	3 000	4 000	5 000	6 000	7 000	8 000	9 000	10 000
kg/cm^2	70	140	210	280	350	420	490	560	630	700
Relative density	0.997	0.992	0.986	0.981	0.977	0.973	0.970	0.967	0.964	0.962

Compressibility effects are commonly ignored on non-critical industrial hydraulic systems working at pressures up to 2 000 psi (140 kg/cm^2), ie the fluid is regarded as incompressible for calculation of the performance of the system. For more critical systems the relative decrease in volume with pressure, and the 'elasticity' introduced in the system, should be regarded as potentially significant with system pressures above 1 000 psi (70 kg/cm^2). The relative decrease in volume likely to be experienced by different fluids at different pressures can readily be estimated from Figs 1.3 and 1.4. More specific information relating to the performance of individual proprietary fluids will be found in Chapters 7, 8 and 9.

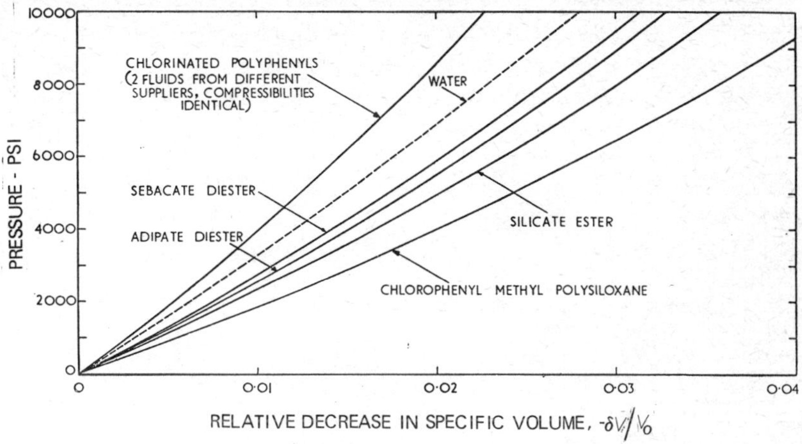

Fig. 1.3 Isentropic compressibilty of miscellaneous fluids at 20 ° C. (NEL Measurements)

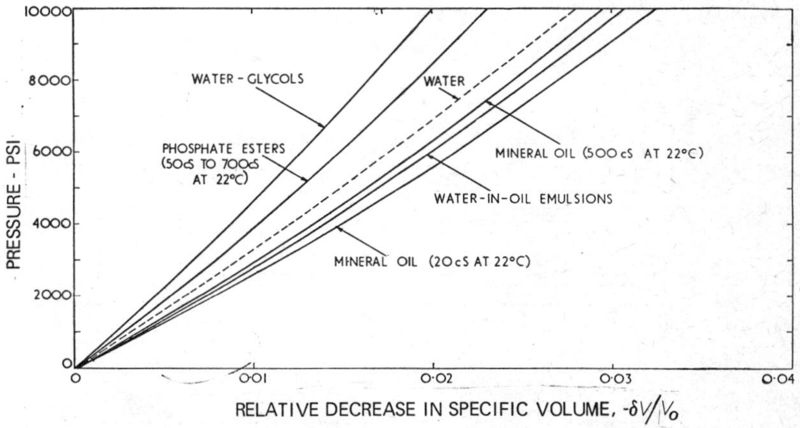

Fig. 1.4 Isentropic cemmpresibility of various hydraulic fluids at 20° C (NEL Measurements)

Fluid compressibility can, however, be markedly increased by the presence of entrained air. Only a relatively small proportion of entrained air can so reduce the bulk modulus of the fluid that the operation of even a low pressure system becomes erratic. It thus becomes strictly necessary to remove any initially entrained air by 'bleeding' the system before it is put into use, and to design the system so that the possibility of subsequent air entrainment is eliminated, or at least minimised.

There is a clear distinction between air in solution in a fluid and entrained air. Most fluids contain a proportion of air in solution – a typical figure being 8 to 9 per cent by volume in the case of hydraulic oils at atmospheric pressure and normal temperatures. Although the solubility of air will vary with pressure and temperature, such changes in the percentage of dissolved air actually present have no marked effect on the compressibility of the fluid, and thus do not require any special treatment.

Effect of Pressure on Viscosity

Fluid viscosity tends to increase with pressure, at constant temperature, this effect usually being most marked at lower temperatures. The rate of increase in viscosity with pressure is a specific property of the fluid concerned, being more marked in the case of mineral oils, for example, than with water-based fluids. Thus, in general figures, the viscosity of a typical mineral oil fluid is approximately doubled, compared with its viscosity at atmospheric pressure and the same temperature, at a pressure of 3000 – 4000 psi (210 – 280 kg/cm²). For a majority of practical applications this effect can generally be ignored, as fluid temperature changes have a more significant effect on fluid viscosity.

Some typical data are given in Tables 1.4 and 1.5 and Figs 1.5 and 1.6; whilst Fig 1.7 shows the temperature-dependent characteristics of

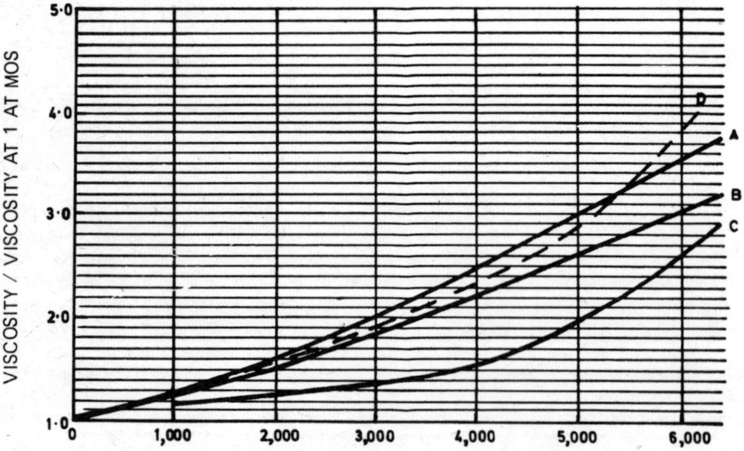

Fig. 1.5 Viscosity pressure characteristics ot 100 degrees for - A-naphthenic base oil. B-mixed base oil. C-paraffenic base oil. D-typical SAE 30 lubricating oil.

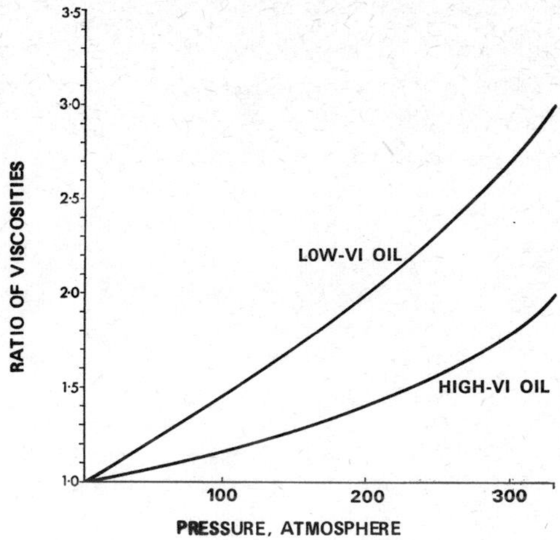

Fig.1.6 Typical Viscosity/pressure characteristics of high VI and low VI oils.

the pressure-viscosity relationship for a typical mineral oil. Pressure has the effect of hastening the thickening of the oil as lower temperatures are reached.

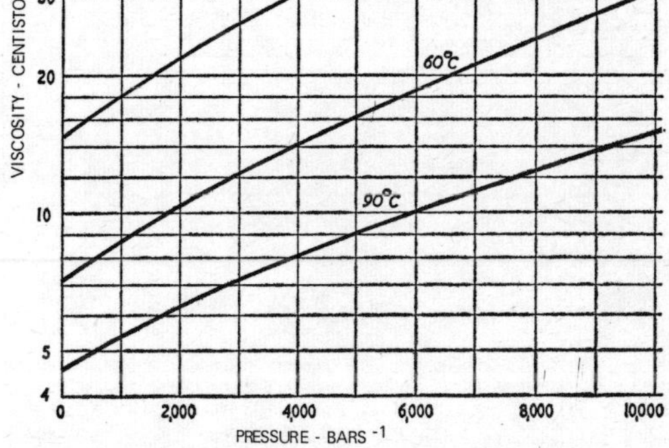

Fig. 1.7.

For any given temperature, the effect on viscosity of raising the fluid from atmospheric to some higher pressure is equivalent to a lowering of the fluid temperature. Raising a typical mineral oil from atmospheric pressure to 3 000 psi (210 kg/cm²), for example, would be equivalent in effect to lowering its temperature from a normal ambient of 68 degrees F (20 degrees C) to about 45 degrees F (7 degrees C), as far as the viscosity is concerned. Thus pressure-viscosity effects, to some extent, offset the loss of viscosity which would be expected from a practical system due to fluid heating under working conditions.

Thermal Expansion and Pressure

The effect of pressure is to reduce the coefficient of thermal expansion of a fluid, very roughly in the order of 0.00005 per cent per 1 000 psi (70 kg/cm²). Thus the actual change in volume of a pressurised fluid with increasing temperature will be less than that predicted by the calculation of thermal effects alone. The actual variation of the coefficient of volumetric expansion with pressure will be specific to a particular fluid, but Fig 1.8 is fairly typical for an oil fluid.

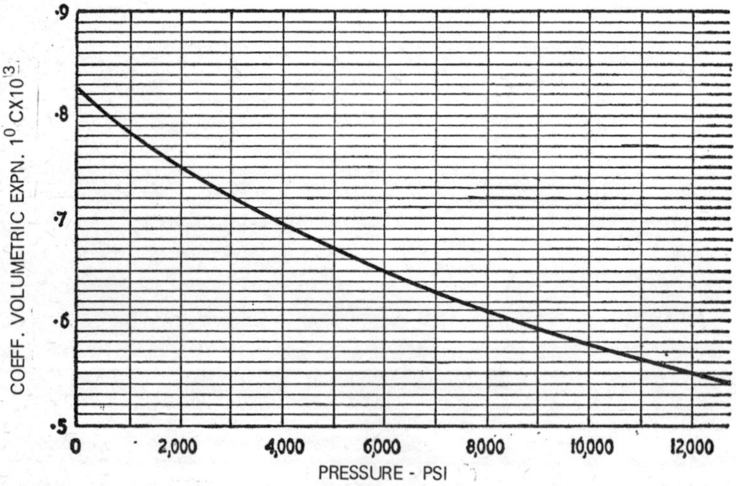

Fig. 1.8 Variation of thermal expansion coefficient with pressure for a typical hydraulic oil.

Methods of Pressurisation

A fluid can be pressurised directly by the application of a force over the whole of one free surface; or indirectly by being pumped – Fig 1.9. In the former case the unit pressure applied (force per unit area) is transmitted through the fluid without loss, in a purely hydrostatic system, and is independent of any loading. In the case of a pump, mechanical energy is converted into fluid energy, using the power input to the pump to change the pressure level of the fluid passing through the pump, and thence into the system. The pump used is almost invariably of positive displacement type and hence the discharge pressure is largely independent of load or 'head'; and in favourable cases pump efficiency can be in excess of 95 per cent.

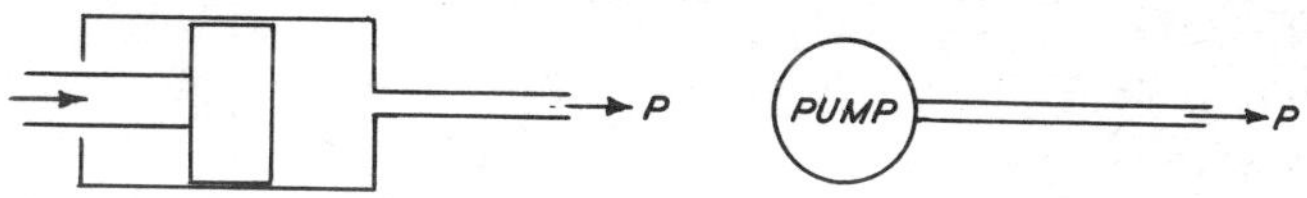

Fig. 1.9.

Centrifugal pumps are not excluded for hydraulic systems. They are employed in modern water-hydraulic systems where large volumes of fluid have to be pumped, and to a limited extent on certain specialised oil-hydraulic systems, eg in the 'Thrustor' unit. The advantages of a centrifugal pump are simplicity of construction and relatively small size for a given capacity. Inherent disadvantages are a relatively low efficiency in smaller sizes, and a pressure output varying with delivery and load.

In any hydrodynamic system involving fluid flow, however, frictional losses are involved, resulting in a pressure drop between the pump and actuator. The maximum overall efficiency likely to be realised (pump input to actuator output) is therefore of the order of 80 per cent. Because of the convenience of the method, efficiency is not all important, particularly when the full output of the pump is not required. Also specialised types of pumps have been developed to meet the power requirements of particular systems.

Pump volumetric efficiency can readily be determined by taking two readings at the intended operating speed, one at zero delivery pressure and one at the required pressure. The difference is a measure of the internal leakage of the pump at the required pressure. This simple test is valid only if the pump is not cavitating under zero delivery pressure. If so, the pump will have to be pressurised to the extent necessary to ensure full displacement.

The mechanical efficiency of a pump can be calculated on a theoretical basis, namely by comparing the calculated input power required to raise the theoretical delivery of the fluid to a given pressure with the actual power input required. Again this is true only if the pump is not cavitating.

The pressurised output from a pump is not necessarily constant. Depending on the type of pump, valve timing (and the number of cylinders in the case of a piston pump), pressure may be derived in a series of pulses – see Fig 1.10. Such cyclic variations usually occur at a frequency equal to the product of the pump speed and number of pump elements involved, which is normally too high to cause jerky movement. In certain applications, however, this pump 'ripple' may be objectionable, or be emphasised by system elasticity, and require damping. Pressure fluctuations may also be produced by variations in delivery rates, and these can be more objectionable than pump ripple.

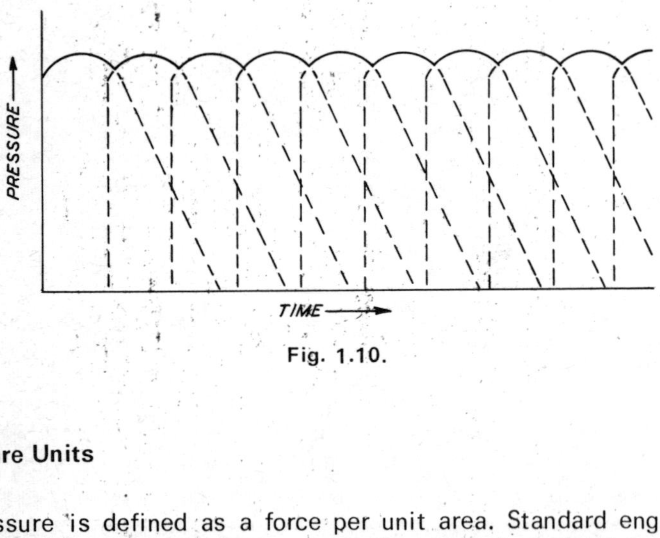

Fig. 1.10.

Pressure Units

Pressure is defined as a force per unit area. Standard engineering units employed in hydraulics are lbf/ square inch (psi) in English units; and kgf/square centimetre (kg/cm^2) in metric units. Pressure may also be expressed as a number relative to normal atmospheric pressure (14.4 psi or 1 kg/cm^2), ie as N At (N times atmospheric pressure, in the appropriate units). This latter designation is little used in engineering hydraulics, except in Continental Europe. Pressure in kg/cm^2 can be read as pressure in atmospheres for most practical purposes.

TABLE 1.6

TYPICAL VISCOSITY – PRESSURE CHARACTERISTICS FOR OILS AT CONSTANT TEMPERATURE
TABLE SHOWS VISCOSITY RELATIVE TO VISCOSITY AT ATMOSPHERIC PRESSURE

	Pressure									
psi	1 000	2 000	3 000	4 000	5 000	6 000	7 000	8 000	9000	10 000
Fluid kg/cm^2	70	140	210	280	350	420	490	560	630	700
SAE 10 Oil	1.12	1.37	1.62	1.89	2.22	2.61	3.40	–	–	–
SAE 30 Oil	1.24	1.57	1.94	2.41	2.75	3.74	–	–	–	–
Paraffin Base	1.2	1.3	1.4	1.6	2.0	2.4	–	3.6	–	–
Naphthenic Base	1.3	1.6	2.0	2.5	3.0	3.5	–	4.8	–	–
Mixed Base	1.3	1.6	1.9	2.2	2.7	3.0	–	4.5	–	–

NOTE: Oils which show the greatest change in viscosity with temperature also show the greatest change in viscosity with pressure – i.e. an improvement in the Viscosity Index of an oil will reduce the viscosity change with pressure.

Very High Pressure Hydraulics

The optimum limit for fluid pressure for actuator systems is of the order of 5 000 – 6 000 psi (350 – 420 kg/cm^2). With lower pressures there is always some advantage to be gained in the way of compactness and power/weight ratio by increasing the system pressure, although this may be outweighed by greater cost because of the need for more precise manufacture of the components. Above about 6 000 psi (420 kg/cm^2), further practical gains in power/weight ratio are no longer realisable, strength requirements become increasingly critical, and compressibility may have a marked effect on the performance of the system. A realistic figure for the upper limit of fluid pressures for hydrostatic power systems is thus probably 8 000 – 10 000 psi (560 – 700 kg/cm^2). Nevertheless, much higher fluid pressures are realised in practice, such as in liquid springs, in chemical processing, metal extrusion, hydraulic forming, pressing and compacting of metal and other powders, and for experimental and pressure testing and auto frettage.

The present limit achieved by fluid pressure devices is of the order of 200 000 psi (14,000 kg/cm^2). It is usual to give pressures above 100 000 psi in kilobars – ie 6.9 kilobars (one kilobar = 1.4504×10^4 psi). Thus 14 kilobars is a nominal ceiling for fluid pressure devices. Above that the subjects are usually solid and pressure is produced by squeezing between anvils.

Mechanically operated pumps are generally impractical for the production of very high pressures, hence intensifiers are almost invariably used for the production of pressures over 50 000 psi. These are fed with a 'low' pressure supply from a suitable pump, this pressure being subject to multiplication by the intensifier. The system may be all-hydraulic, when the primary or supply pressure may be anything up to about 16 000 psi (1.125 kg/cm^2) to yield a final pressure of 200 000 psi (14 kilobars); or air-hydraulic. In the latter case input power is taken from a compressed air supply (100 psi in the case of the Airhydropump), and practical output pressures of the order of 150 000 psi (10 kilobars) are realisable.

The fluids used in very high pressure systems present a separate problem. Ordinary mineral oils become so viscous as to be virtually solid at pressures of the order of 70 000 psi (4,900 kg/cm^2). The use of viscosity index improvers enable oil fluids to be used satisfactorily at pressures up to about 100 000 psi (6.9 kilobars). Castor oil can also be used up to similar pressures, and castor-base fluids up to about 150 000 psi (10 kilobars). Water-glycol solutions and synthetic fluids with high bulk modulii can be used for pressures up to 200 000 psi (14 kilobars) – See also Table 1.7. Glycerine is also an excellent fluid

TABLE 1.7

TYPICAL PRESSURE LIMITS FOR FLUIDS

Fluid	Maximum Practical Working Pressure		
	psi	kg/cm^2	kilobars
Mineral Oils	60 000	4 200	4
Mineral Oils with VI improvers	100 000	7 000	7
Castor Oil	100 000	7 000	7
Castor-Base Fluids	150 000	10 500	10
Di-Methyl-Hexyl Sebacate	150 000	10 500	10
Water-Glycol Solutions	200 000	14 000	14
Glycerine	200 000	14 000	14

TABLE 1.8

DENSITY OF GLYCERINE AT CONSTANT TEMPERATURE
(density in grams/cc)

Pressure psi kg/cm^2	0	2 000	4 000	6000	8 000	10 000
Density	1.26	1.27	1.28	1.29	1.29	1.30

for very high pressure working since it has extremely low compressibility – see Table 1.8. It may be used on its own, or as a glycerine-water solution. The principle disadvantage of glycerine is that it tends to cause rusting – a similar limitation applying to other water-base solutions, which may require inhibitors.

The compressibility of the fluid is particularly significant for very high pressure working, for the greater the compressibility, the lower the volumetric efficiency of the intensifier, ie the greater the power input required for a given output. The bulk of this power loss is stored in the compressed fluid, and will be released when the pressure restraint is removed. This can represent a hazard in the case of the high pressure system failure, or where the very high fluid pressure is being used for a destruction test.

2. HYDROSTATIC THEORY AND PRACTICE

In a fluid at rest no relative movement is experienced between fluid layers and thus the effects of viscosity can be ignored. The pressure in a static fluid is scalar quantity. That is to say it is a quantity without any direction; or equally the pressure is the same in all directions. Once a surface is immersed, or in intimate contact with the fluid surface, the pressure on that surface is expressed by a definite vector in a direction normal to the surface. The magnitude of this vector is given by P.A, where P is the pressure and A the area of the surface.

This is an arithmetical summary of Pascal's laws of hydrostatics stated originally in 1948:-

(i) A static fluid exerts pressure at right angles to any plane surface containing it.

(ii) The pressure at any point in a static fluid is the same in every direction.

(iii) The pressure in a static fluid is the same for all points, provided the weight of the fluid is neglected.

Pressure in a fluid at rest will increase with increasing depth below the free surface (or effectively with the height of fluid above the position under consideration.) For a fluid of constant specific weight the pressure change (Δp) can be expressed in the simple form

$\Delta p = w\,h$

where w = specific weight of fluid

h = height or vertical distance between the two points under consideration

Position may also be defined relative to an arbitrary datum, eg z in Fig 2.1. This 'position height' is called the *potential head,* with respect to that datum. The quantity p/w is called the *pressure head;* and the sum of the two, the *piezometric head.* Thus:-

peizometric head = $p/w + z$

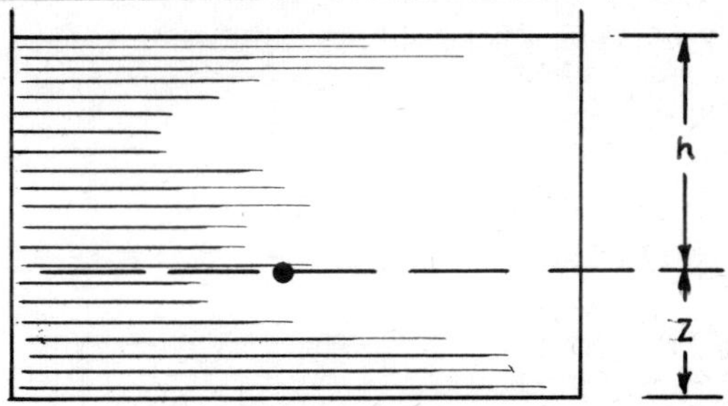

Fig. 2.1

It follows that a change of position producing an increase in pressure head will produce a corresponding decrease in potential head, and vice versa. Thus the piezometric head of a fluid at rest is constant. This is a special case of the Bernouilli equation where the velocity terms are zero. In fact, hydrodynamic equations are directly applicable to fluid statics by rendering all the terms involving fluid motion as zero. Conversely, the application of hydrostatic principles for the transmission of force through a fluid must take into account velocity terms if fluid movement is involved, although in practice these may be small enough to be negligible. This applies generally when flow rates are small, when the fluid performs a 'passive' role in the transmission of pressure and the small amount of fluid movement involved has no significant effect on the performance of the system.

In practice, the majority of straightforward hydraulic systems are hydrostatic in character, transmitting power by pressure energy. This is re-converted into mechanical output by an actuator — or, more correctly, a hydrostatic machine — the whole forming what is called a hydraulic system. The performance of such a system is largely unaffected by the fluid characteristics other than its compressibility. However, once appreciable movement is introduced, fluid friction, mechanical friction, inertia and other velocity terms tend to become increasingly significant and the system is operating partly as a hydrostatic one and partly as a hydrodynamic one. Thus high flow rate systems may be essentially hydrodynamic in function although they may still be employed with hydrostatic machines. Estimating performance on hydrostatic principles will then yield erroneous (over-optimistic) results. Also once hydrodynamic terms are introduced, performance is greatly affected by the fluid characteristics, notably the viscosity, governing the loss of pressure energy.

At the same time a clear distinction can be drawn by systems involving hydrodynamic reactions or dynamic forces, but are basically hydrostatic systems; and those which are designed to utilise the kinetic energy of the fluid medium.The latter are known as *hydrokinetic* systems, and the converter which transforms this fluid energy into mechanical output, a hydrokinetic machine. This may also, and usually does, involve hydrodynamic reactions. Thus 'hydrostatic' and 'hydrokinetic' are basically different forms of power transmission,by pressure energy and kinetic energy, respectively. Hydrodynamic functions are related to fluid movement and can be common to both types of systems; or a combination of such systems.

As examples: the hydraulic press is a typical hydrostatic system with small fluid movement and generally negligible hydrodynamic (velocity) effects. A hydraulic cylinder is a hydrostatic machine which works on exactly the same principle (hydrostatics), although its actual performance in a working system may be modified by pressure losses resulting from hydrodynamic effects. Knowing these losses, its performance can be calculated directly as behaving like a simple hydrostatic machine.

At this point it is desirable to mention hydrostatic and hydrokinetic *transmissions* in order to clarify these terms and avoid possible confusion. Both are basically rotary systems, and are described in more detail in Chapter 5. It is merely necessary to state here that –

A *hydrostatic transmission* is essentially a hydrodynamic device, utilising the pressure energy in a fluid developed by a pump and converting this by hydrostatic reaction with no significant change in fluid velocity. The energy is transmitted by a relatively small flow of fluid at high pressure.

A *hydrokinetic transmission* is a kinetic energy converter, where energy is derived from the change in velocity of the fluid. It operates with comparatively large flows at low pressures.

Force Multiplication

Pascal's laws may be re-stated – 'a pressure exerted on a fluid in a closed system is transmitted equally to all parts of the system and acts perpendicularly to all surfaces in contact with the fluid'. The assumption that the weight of the fluid can be neglected for hydrostatic applications is valid since the potential head involved is relatively very small compared with the applied pressure.

The classic example of the practical application of this basic hydrostatic principle is the Brahmah press, shown diagrammatically in Fig 2.2. A force F1 is applied to a piston (1) of cross sectional area A1, developing a pressure of (F1/A1) throughout the fluid. This pressure is transmitted through the fluid equally in all directions so that the cylinder walls, walls of the interconnecting pipe, and the bottom of the piston in cylinder (2) are all subjected to a pressure of (F1/A1). The force exerted in piston (2) is therefore (F1/A1) x A2, or

$$F2 = F1 \times \frac{A2}{A1}$$

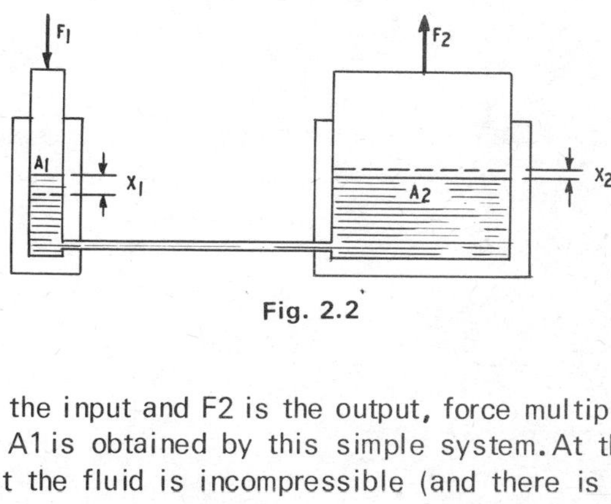

Fig. 2.2

Thus if F1 is the input and F2 is the output, force multiplication in the ratio of A2 : A1 is obtained by this simple system. At the same time, assuming that the fluid is incompressible (and there is no leakage of fluid past the pistons), the distance travelled by the respective pistons is in inverse ratio to the piston areas, viz

$$\frac{X_1}{X_2} = \frac{A1}{A2}$$

This is a true hydrostatic system for, although a small displacement of fluid is involved, velocity terms can be neglected. In a practical system, however, the performance may be modified by the weight of the two pistons and the friction of the piston seals (necessary to prevent escape of fluid past the pistons when pressurised).

Thus the true pressure developed in the fluid (see Fig 2.3) is given by

$$P = \frac{4\,(F_{in} + W_1 - f_1)}{\pi\, D_1^{\,2}}$$

where W_1 is the weight of piston (1)

F_1 is the seal friction

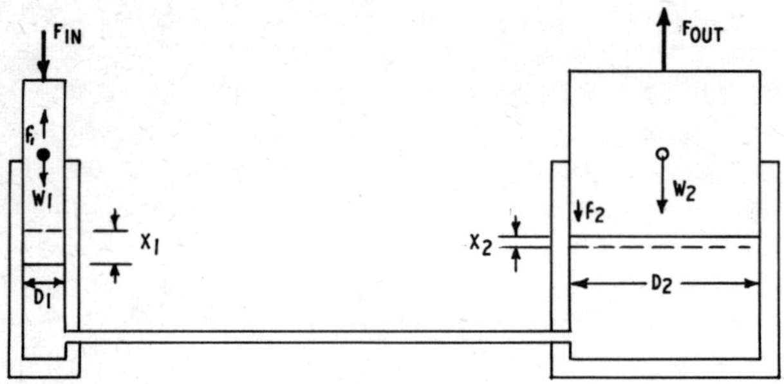

Fig. 2.3

This pressure acting on the bottom of piston (2) produces an output force –

$$F_{out} = P \times \frac{\pi D_2^2}{4} - W_2 - f_2$$

$$= \left[F_{in} + W_1 - f_1\right]\frac{D_2^2}{D_1^2} - (W_2 - f_2)$$

or if piston weight is neglected

$$F_{out} = (F_{in} - f_1)\,\frac{D_2^2}{D_1^2} - f_2$$

The respective piston travels are unaltered, provided the fluid is incompressible and there is no leakage past the seals, thus –

$$X_2 = X_1\left[\frac{D_1^2}{D_2^2}\right]$$

These equations thus fully define the performance of a basic hydrostatic system of force multiplication for pressure ranges over which compressibility effects can be neglected. It will be seen that this performance is independent of any other fluid property, eg fluid viscosity has no affect other than that of controlling leakage and possibly modifying seal friction. Thus an oil would be favoured to water in such systems because its higher viscosity would reduce the possibility of leakage, and its lubricating properties help reduce seal friction.

The original Brahmah press of 1795 specified a ¼ in diameter input piston and a 12 in diameter output piston, giving a force multiplication of 2304 – Fig 2.4. The lever was specified to provide an input of up to 20 cwt. Thus yielding an output force of 2304 tons! The practical success of this press, incidentally, was dependent almost entirely on the development of a 'self-sealing' leather piston seal by Maudsley, permitting the high pressure to be maintained without leakage and providing easy return of the pistons when pressure was relieved.

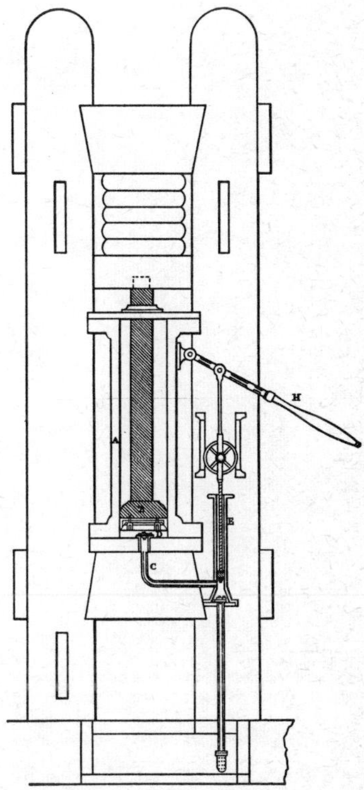

Fig. 2.4 Original drawing of Bramah's press. A - iron cylinder; B - piston; C - tube; E - pump or injector; H - lever.

A common modern example of the force multiplier is the hydraulic brake system applied to automobiles – Fig 2.5. In this case multiplication is achieved mechanically via a simple lever system, since it is inconvenient to employ large output cylinders. Also the movement required by the output cylinders is quite small, and the master cylinder has to supply fluid to four separate output cylinders simultaneously. Slave piston movement is thus kept to a minimum practical value to reduce the stroke requirements of the master cylinder. This enables a high degree of mechanical input multiplication to be achieved without excessive input movements. A system pressure of 600 to 800 psi (40 to 60 kg/cm^2) can then be achieved with a maximum input (pedal) pressure of the order of 100 pounds with a master cylinder diameter of 1 inch (25 mm). No force multiplication occurs in the hydraulic circuit; in fact, rather the reverse since the slave cylinders are normally of smaller diameter than the master cylinder.

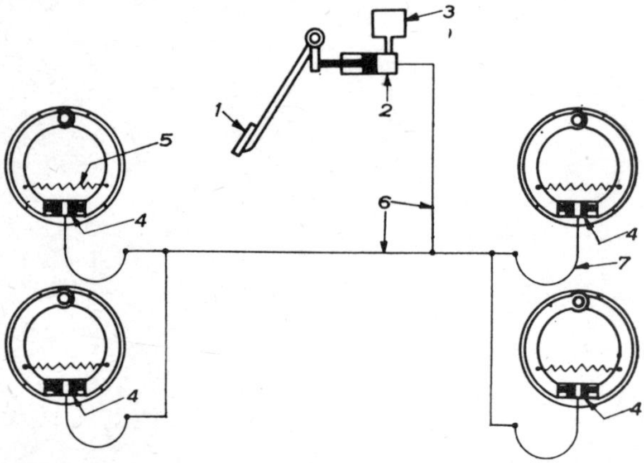

Fig. 2.5 1.Brake pedal. 2.Master cylinder. 3.Reservoir. 4.Slave cylinders. 5.Return springs. 6.Metal pipes. 7.Flexible hoses.

The hydraulic circuit is of simple one-way type, with spring return of the output cylinders. No control valves are necessary since fluid flow is controlled entirely by master piston movement and the spring return. The addition of a reservoir is necessary, however, both to ensure that the system always remains filled, and to accommodate changes in fluid volume with temperature.

As long as the system remains completely filled with fluid and remains free from leaks, force transfer should approach 100 per cent efficiency – it being, in fact, a force transfer system rather than a force multiplier, as far as the hydraulics are concerned. The main requirements of the fluid used are that it should have an optimum viscosity to promote efficient sealing at all likely ambient temperatures and promote adequate lubrication of the piston seals. It should also have high chemical and physical stability since it is in contact with air in the reservoir. Equally, of course, it must be compatible with the piston seals.

There are numerous other examples of force multipliers with manual input where the multiplication of the input force is achieved mechanically and the hydraulics perform primarily as a force transfer system – such as so-called 'servo' or servo-assisted systems. This is practical and expedient where mechanical drive (with multiplication) can easily be applied for linear drive of the master or 'transmitter' cylinder. The latter can be duplicated, if necessary, to provide hydraulic force transmission in either direction as in the case of telemotor systems working with a simple closed circuit.

In a majority of cases of force-transmitting systems the transmitter is a circulating pump, working in a closed circuit. The input is then determined by the rated pump performance, ie pressure it can produce and quantity of fluid it can deliver. Choice of fluid then becomes more critical, particularly as regards its viscosity and stability.

Viscosity limits are normally set by lubrication requirements, the pump being the most critical part in the system. A viscosity which is too low will increase internal leakage, and may provide inadequate lubrication. A viscosity which is too high will increase power losses in pumping, possibly produce overheating of the fluid and even cavitation. The fluid selected must, therefore, have an optimum working viscosity for the particular pump involved. Emphasis should be placed on the *working* viscosity as this must take into account both ambient temperature range of heating experienced by the fluid during working of the system.

Stability requirements are also important since the fluid is circulated and thus subject to more complex working conditions which can promote degradation. This subject is dealt with in detail in later chapters.

Pump Power

Where the pump forms the transmitter in a force transfer system, the output power required can be derived from basic hydrostatic principles. As a typical example, take the case of a pump required for powering a hydraulic press – Fig 2.6. – and assume that velocity components are negligible.

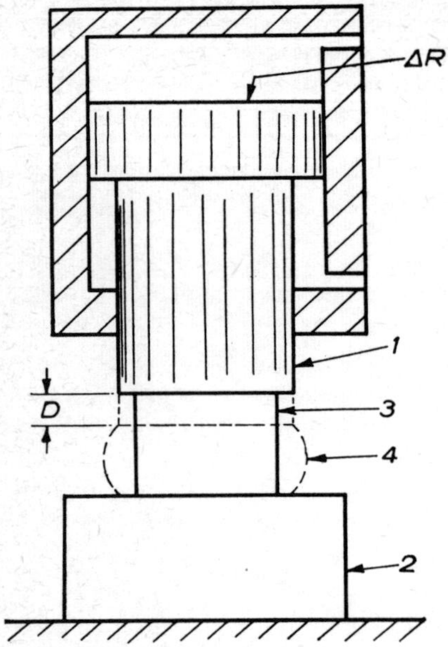

Fig. 2.6 1.Ram. 2.Anvil. 3.Part to be compressed. 4.Part after compression.

The press ram is 'sized' on the basis of the output force required and a practical figure for the fluid pressure to be employed (based largely on the performance of available pumps). Obviously the higher the pressure the smaller the ram can be for a given output force and so the initial solution must be a compromise between size and fluid pressure.

$$\text{Ram diameter } D = \sqrt{\frac{4F}{\pi P}} = 1.13\sqrt{\frac{F}{P}}$$

in consistent units

Strictly speaking this calculation should also take into account the weight of the ram (which may be positive or negative, relative to the force output; or effectively zero in the case of a horizontal ram); and the friction of the ram seals.

$$\text{Ram diameter} = 1.13\sqrt{\frac{F \pm W_r - f_{r1} - f_{r2}}{P}}$$

It is then necessary to decide on the speed of working e.g. a stroke of L in time t seconds, or a speed of L/t. The size of pump, i.e. delivery required, then follows from

$$\text{delivery (Q)} = \frac{D^2}{4} \times \frac{L}{t}$$

$$\text{or} = \left[\frac{F \pm W_r - f_{r1} \quad f_{r2}}{P}\right] \times \frac{L}{t}$$

in consistent units

This defines the output requirements of the pump – i.e. delivery pressure (P) and delivery output (Q). The pump (output) power follows as –

$$\text{HP (pump)} = \frac{F \times L}{N \times t}$$

where N is the appropriate HP constant for the units employed,

$$\text{HP (pump)} = \frac{\text{F (pounds)} \times \text{L (inches)}}{6600 \times \text{t (secs)}}$$

$$\text{HP (pump)} = \frac{\text{F (tons)} \times \text{L (inches)}}{2.95 \times \text{t (secs)}}$$

$$\text{HP (pump)} = \frac{\text{F (kilograms)} \times \text{L (centimetres)}}{3700 \times \text{t (secs)}}$$

The significance of the speed of operation of the ram is obvious. For the same output force the pump power required is halved by doubling the time for a given stroke. In practice very considerable savings in pump power can often be realised by reducing speed substantially at maximum load.

Although all ram forces are determined on the basic hydrostatic principle, actual performance may be modified by back pressure effects and velocity components. The subject is further dealt with in Chapter 4.

TABLE 2.1
DISPLACEMENT TABLE (UNIVERSAL)

Bore	STROKE OR LENGTH									
	1	2	3	4	5	6	7	8	9	10
.5	.196	.393	.589	.786	.972	1.058	1.375	1.571	1.767	1.964
.6	.283	.566	.849	1.131	1.414	1.697	1.980	2.262	2.546	2.827
.7	.385	.770	1.155	1.540	1.925	2.310	2.694	3.080	3.464	3.849
¾	.442	.884	1.326	1.768	2.209	2.652	3.094	3.536	3.478	4.418
.8	.503	1.006	1.509	2.012	2.514	3.017	3.520	4.023	4.525	5.027
.9	.636	1.272	1.908	2.544	3.181	3.817	4.453	5.089	5.725	6.362
1.0	.785	1.570	2.355	3.140	3.92	4.710	5.295	6.280	7.065	7.854
1.1	.950	1.900	2.851	3.801	4.752	5.702	6.652	7.602	8.552	9.503
1.2	1.131	2.262	3.393	4.524	5.655	6.786	7.917	9.048	10.179	11.310
1¼	1.227	2.454	3.681	4.908	6.14	7.262	8.589	9.816	11.043	12.27
1.3	1.327	2.654	3.982	5.309	6.637	7.964	9.291	10.618	11.946	13.273
1.4	1.539	3.079	4.618	6.158	7.697	9.236	10.775	12.315	13.854	15.394
1.5	1.767	3.534	5.301	7.068	8.825	10.602	12.369	14.136	15.903	17.67
1.6	2.011	4.022	6.032	8.043	10.053	12.064	14.074	16.085	18.095	20.106
1.7	2.270	4.540	6.810	9.080	11.349	13.619	15.889	18.159	20.429	22.698
1¾	2.405	4.810	7.215	9.620	12.025	14.430	16.835	19.240	21.645	24.05
1.8	2.545	5.090	7.635	10.180	12.724	15.269	17.813	20.358	22.903	25.447
1.9	2.835	5.670	8.507	11.342	14.177	17.012	19.848	22.683	25.518	28.353
2.0	3.142	6.284	9.426	12.568	15.708	18.852	21.994	25.136	28.278	31.416
2.1	3.463	6.926	10.389	13.852	17.315	20.778	24.241	27.704	31.167	34.63
2.2	3.801	7.602	11.403	15.204	19.005	22.806	26.607	30.408	34.209	38.01
2¼	3.976	7.952	11.928	15.904	19.83	23.856	27.832	31.808	35.784	39.76
2.3	4.155	8.310	12.465	16.620	20.775	24.930	29.085	33.240	37.395	41.55
2.4	4.524	9.048	13.572	18.096	22.620	27.144	31.668	36.192	37.716	45.24
2.5	4.909	9.818	14.727	19.636	24.545	29.454	34.363	39.272	44.181	49.09
2.6	5.309	10.618	15.927	21.236	26.545	31.854	37.163	42.472	47.781	53.09
2.7	5.726	11.452	17.178	22.904	28.630	34.356	40.082	45.808	51.534	57.26
2¾	5.940	11.880	17.820	23.760	29.70	35.640	41.580	47.520	52.460	59.40
2.8	6.157	12.314	18.471	24.628	30.785	36.942	43.099	49.256	55.413	61.57
2.9	6.605	13.210	19.815	26.420	33.025	39.630	46.235	52.840	59.445	66.05
3.0	7.069	14.138	21.207	28.278	35.24	42.414	49.483	56.556	63.621	70.67
3.1	7.548	15.096	22.644	30.192	37.740	45.288	52.836	60.384	67.932	75.48
3.2	8.042	16.084	24.126	32.168	40.210	48.252	56.294	64.336	72.378	80.42
3¼	8.296	16.592	24.888	33.184	41.48	49.776	58.072	66.368	74.664	82.96
3.3	8.553	17.106	25.659	34.212	42.765	51.318	59.871	68.424	76.977	85.53
3.4	9.079	18.158	27.237	36.316	45.395	54.474	63.553	72.632	81.711	90.79
3.5	9.621	19.242	28.863	38.484	48.11	57.726	67.347	76.968	86.589	96.21
3.6	10.18	20.36	30.54	40.72	50.90	61.08	71.26	81.44	91.62	101.79
3.7	10.75	21.50	32.29	43.00	53.76	64.51	75.26	86.01	96.76	107.52
3¾	11.045	22.090	33.135	44.180	55.23	66.270	77.315	88.36	99.405	110.45
3.8	11.34	22.68	34.02	45.36	56.71	68.05	79.39	90.73	102.07	113.41
3.9	11.95	23.89	35.84	47.78	59.73	71.67	83.62	95.56	107.51	119.46
4.0	12.566	25.132	37.698	50.264	62.83	75.396	87.962	100.53	113.09	125.66
4.1	13.20	26.40	39.61	52.81	66.02	79.22	92.42	105.62	118.82	132.03
4.2	13.86	27.71	40.57	54.43	69.29	83.14	96.99	110.84	124.70	138.55
4¼	14.186	28.372	42.558	56.744	70.93	85.116	99.302	113.49	127.67	141.9
4.3	14.52	29.04	43.56	58.08	72.61	87.13	101.65	116.17	130.69	145.22
4.4	14.21	30.42	45.63	60.83	76.03	91.24	106.45	121.66	136.86	152.05
4.5	15.904	31.808	47.712	63.616	79.52	95.424	111.33	127.23	143.14	159.0

These tables give theoretical displacement or swept volume in any series of consistent units i.e., Bore and stroke in inches, gives displacement in cubic inches.
Bore and stroke in centimetres, gives displacement in cubic centimetres.
For intermediate solutions values of *strokes* may be interpolates.
Intermediate values of bore may *not* be interpolated.

TABLE 2.1 (continued)

DISPLACEMENT TABLE (UNIVERSAL)

Bore	STROKE OR LENGTH									
	1	2	3	4	5	6	7	8	9	10
4.6	16.62	33.24	49.86	66.48	83.10	99.72	116.34	132.96	149.58	166.19
4.7	17.350	34.700	52.050	69.400	86.750	104.10	121.45	138.80	156.15	173.50
4¾	17.721	35.442	53.163	70.884	88.61	106.33	124.05	141.77	159.49	177.2
4.8	18.100	36.192	54.290	72.380	90.48	108.58	126.67	144.76	162.86	180.96
4.9	18.86	37.72	56.58	75.44	94.29	113.15	132.01	150.87	169.73	188.58
5.0	19.635	39.270	58.905	78.540	98.18	117.81	137.45	157.08	176.72	196.35
5.1	20.43	40.86	61.29	81.72	102.14	122.58	143.01	163.44	183.87	204.28
5.2	21.24	42.48	63.72	84.96	106.19	127.43	148.66	169.90	190.14	212.37
5¼	21.648	43.296	64.944	86.592	108.24	129.89	151.54	173.18	194.83	216.48
5.3	22.06	44.12	66.18	88.24	110.31	132.36	154.42	176.48	198.54	220.62
5.4	22.90	45.80	68.70	91.60	114.51	137.41	160.31	183.21	206.11	229.02
5.5	23.758	47.516	71.274	95.032	118.79	142.55	166.31	190.06	213.82	237.58
5.6	24.63	49.26	73.89	98.52	123.15	147.78	172.31	197.04	231.67	246.30
5.7	25.52	51.04	76.56	102.08	127.59	153.11	178.63	204.15	229.67	255.18
5¾	25.967	51.934	77.901	103.87	129.84	155.80	181.72	207.74	233.70	259.67
5.8	26.42	52.84	79.26	105.68	132.10	158.52	184.94	211.36	237.78	264.21
5.9	27.34	54.68	82.02	109.36	136.70	164.04	191.38	218.72	246.06	273.40
6.0	28.274	56.548	84.822	113.10	141.37	169.64	197.89	226.20	254.47	282.74
6.1	29.23	58.46	87.68	116.91	146.13	175.36	204.59	233.81	263.03	292.25
6.2	30.19	60.38	90.57	120.76	150.95	181.14	211.33	241.52	271.71	301.91
6¼	30.680	61.360	92.040	122.72	153.40	184.08	214.76	245.44	276.12	306.8
6.3	31.17	62.34	93.51	124.68	155.8	187.02	218.19	249.36	280.53	311.7
6.4	32.17	64.34	96.51	128.68	160.85	193.02	225.19	257.36	289.53	321.70
6.5	33.183	66.366	99.549	132.73	165.92	199.10	232.28	265.46	298.65	331.83
6.6	34.21	68.42	102.63	136.84	171.06	205.26	239.47	273.68	307.89	342.12
6.7	35.26	70.52	105.78	141.03	176.28	212.54	247.80	283.05	308.31	352.57
6¾	35.785	71.570	107.36	143.14	178.93	214.72	250.49	286.28	322.08	357.85
6.8	36.32	72.64	108.96	145.28	181.59	217.91	254.23	290.54	326.86	363.17
6.9	37.39	74.78	112.18	149.57	186.97	224.36	261.76	299.15	336.64	373.93
7.0	38.485	76.970	115.46	153.94	192.43	230.92	269.39	307.88	346.38	384.85
7.1	39.59	79.18	118.77	158.36	197.96	237.55	277.14	316.73	356.32	395.92
7.2	40.72	81.43	122.15	162.87	203.58	244.30	285.01	325.73	366.44	407.15
7¼	41.282	82.564	123.85	165.13	206.41	247.70	288.97	330.26	371.55	412.8
7.3	41.85	83.71	125.56	167.41	209.27	251.12	292.98	334.83	376.69	418.54
7.4	43.01	86.02	129.03	172.04	215.05	258.06	301.07	344.08	387.09	430.09
7.5	44.179	88.358	132.54	176.72	220.85	265.08	309.25	353.44	397.62	441.79
7.6	45.37	90.73	136.10	181.46	226.82	272.19	317.55	362.92	408.28	453.65
7.7	46.57	93.13	139.76	186.26	232.83	279.39	325.96	372.52	419.09	465.66
7¾	47.173	94.346	141.52	198.70	235.87	283.04	330.21	397.40	424.56	471.73
7.8	47.78	95.57	143.35	191.14	238.92	286.71	334.49	382.28	430.06	477.84
7.9	49.02	98.03	147.05	196.06	245.08	294.09	343.11	392.13	441.15	490.17
8.0	50.265	100.53	150.80	201.06	251.33	301.60	351.86	402.12	452.40	502.7
8.5	56.745	113.49	170.24	226.98	283.73	340.48	397.22	453.96	510.72	567.5
9.0	63.617	127.23	190.83	254.46	318.09	318.66	445.32	508.92	572.49	606.2
9.5	70.882	141.76	212.65	283.52	359.41	425.30	496.17	567.04	637.95	708.8
10.0	78.540	157.08	235.62	314.16	392.70	471.24	549.78	628.32	706.86	785.4
10.5	86.590	173.18	259.77	346.36	432.95	519.54	606.13	692.72	779.31	865.9
11.0	95.033	190.07	285.10	380.14	475.16	570.20	665.23	760.28	855.30	950.3
11.5	103.87	207.74	311.61	415.48	519.35	623.22	727.09	830.96	934.83	1038.7
12.0	113.10	226.20	339.30	452.40	565.50	678.60	791.70	904.80	1017.90	1131.0

TABLE 2.1 (continued)

DISPLACEMENT TABLE (UNIVERSAL)

Bore	STROKE OR LENGTH									
	10	20	30	40	50	60	70	80	90	100
13	1,327.3	2,654	3,982	5,309	6,637	7,964	9,291	10,618	11,946	13,273.2
14	1,539.4	3,079	4,618	6,158	7,697	9,236	10,775	12,315	13,854	15,393.8
15	1,767.2	3,534	5,301	7,068	8,835	10,602	12,369	14,136	15,903	17,671.5
16	2,010.6	4,021	6,032	8,043	10,053	12,064	14,074	16,085	18,095	20,106.2
17	2,269.8	4,540	6,810	9,080	11,349	13,619	15,889	18,159	20,429	22,698.0
18	2,544.7	5,090	7,635	10,180	12,724	15,269	17,813	20,358	22,903	25,446.9
19	2,835.3	5,670	8,507	11,342	14,177	17,012	19,848	22,683	25,518	28,352.9
20	3,141.6	6,284	9,426	12,568	15,708	18,852	21,994	25,136	28,278	31,415.9
21	3,463.6	6.926	10,389	13,852	17,315	20,778	24,241	27.704	31,167	34,636.1
22	3,801.3	7.602	11,403	15,204	19,005	22,806	26,607	30,408	34,209	38,013.4
23	4,154.8	8,310	12,465	16,620	20,775	24,930	29,085	33,240	37.395	41,547.7
24	4,523.9	9.048	13,572	18,096	22,620	27,144	31,668	36.192	40,713	45,238.9
25	4,908.7	9,818	14,727	19,636	24.545	29,454	34,363	39,272	44,181	49,087.4
26	5,309.3	10,618	15,927	21,236	26,545	31,554	37.163	42,472	47,781	53,092.9
27	5,725.6	11,452	17,178	22,904	28,630	34,356	40,082	45,808	51,534	57,255.6
28	6,157.5	12,314	18,471	24,628	30,785	36,942	43,099	49,256	55,413	61,575.3
29	6,605.2	13,210	19,815	26,420	33,025	39,630	46,235	52,840	59,445	66,052.1
30	7,068.6	14,138	21,207	28,278	35,235	42,414	49,483	56,556	63,621	70,686.0
31	7,547.7	15,096	22,644	30,192	57,740	45,288	52,836	60,384	67,932	75,476.9
32	8,042.5	16,084	24,126	32,168	40,210	48,252	56.294	64,336	72,328	80,424.9
33	8,553.0	17,106	25,659	34,212	42,765	51,318	59,871	68,424	76,977	85,530.0
34	9,079.2	18,158	27,237	36,316	45,395	54,474	63,553	72,632	81,711	90,792.1
35	9.621.1	19.242	28,863	38,484	48,107	57,726	67,347	76,968	86,589	96,211.4
36	10.179	20,360	30,540	40,720	50,900	61,080	71,260	81,440	91,620	101,788
37	10,752	21,500	32,250	43,000	53,760	64,510	75,260	86,010	96,760	107,521
38	11,341	22,680	34,020	45,360	56,710	68,050	79,390	90,730	102,070	113,412
39	11,946	23,890	35,840	47,780	59,730	71,670	83,620	95,560	107,510	119,459
40	12,566	25,132	37,698	50,264	62,830	75,396	87,962	100,530	113,090	125,664
41	13,203	26,400	39,610	52,810	66,020	79,220	92,420	105,620	118,820	132,026
42	13,855	27,710	40,570	54,430	69,280	83,140	96,990	110,840	124,700	138,545
43	14,522	29,040	43,560	58,080	72,510	87,130	101,650	116,170	130,690	145,220
44	15,205	30,420	45,630	60,830	76,030	91,240	106,450	121,660	136,860	152,053
45	15,904	31,808	47,712	63,616	79,520	95,424	111,330	127,230	143,140	159,043
46	16,619	33,240	49,860	66,480	83,100	99,720	116,340	132,960	149,580	166,191
47	17,350	34,700	52,050	69,400	86,750	104,100	121,450	138,800	156,150	173,495
48	18,010	36,192	54,288	72,380	90,480	108,576	126,672	144,760	162,864	180,956
49	18,858	37,720	56,580	75,440	94,290	113,150	132,010	150,870	169,730	188,575
50	19,635	39,270	58,905	78,540	98,175	117,810	137,450	157,080	176,720	196,350

3 HYDRODYNAMIC THEORY AND PRACTICE

Fluid flow is *one-dimensional* when all the fluid and flow parameters are constant throughout any cross section normal to the flow, although changes in flow velocity may occur from section to section. Fluid flow is *two-dimensional* where either, or both, the fluid and flow parameters have gradients in two directions in space; and flow is *three-dimensional* where these gradients exist in random directions and can be defined with reference to the x, y and z axes in a rectangular system of co-ordinates.

In practice the flow of real fluids cannot be one-dimensional since the velocity of flow at a boundary will be zero relative to the boundary. For simplicity, however, analysis is usually based on one-dimensional flow, implying average values of the fluid and flow parameters. This can give quite satisfactory results for engineering hydraulic calculations. In the case of flow through ducts and pipes, flow is actually normally three-dimensional, although for more detailed analysis it can be assumed symmetrical about the duct axis – i.e. two-dimensional flow with axial and radial gradients.

One-dimensional flow in a streamline direction may be considered in terms of flow through a stream tube, i.e. a flow passage bounded by streamlines so that no flow passes through the walls of the stream tube – Fig.3.1. Velocity vectors are tangent to streamlines at all points, thus no fluid can cross a streamline, and no streamline can intersect or cross another streamline. Flow through pipes conforms to this concept, the pipe walls forming the boundary of a stream tube.

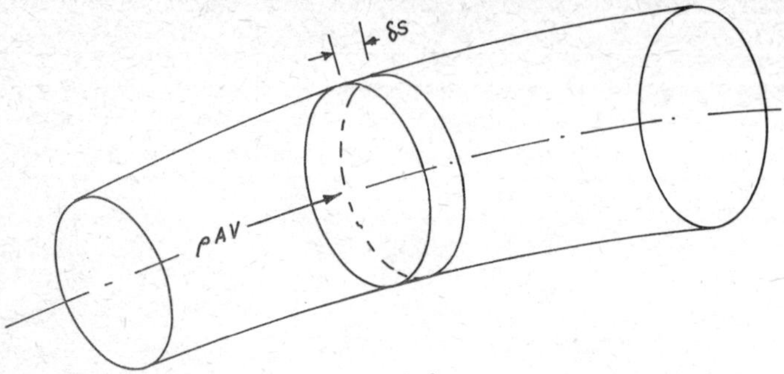

Fig. 3.1

For unsteady flow conditions

$$\frac{\delta(\rho A.V)}{\delta S} + \frac{\delta(\rho A)}{\delta t} = 0$$

For steady flow, the time component is eliminated and so the flow condition is simply

$$\frac{\delta(\rho A.V)}{\delta S} = 0$$

This can be expressed in the form

$$\rho A.V. = \underline{m}$$

where $\underline{m}$ is a constant mass flow rate

and V is the average velocity through the cross section A of the stream tube

This leads directly to the fact that for incompressible flow, where density (ρ) is constant, the volumetric flow rate (Q) is derived directly as

$$Q = V.A. = V_1 A_1 = V_2 A_2 \text{ etc}$$

If the cross sectional area is constant ($A_1 = A_2$, etc.) the flow rate is constant for a given velocity.

The differential form of the stream tube steady flow equation is

$$\frac{1}{V}\frac{dV}{dS} + \frac{1}{A}\frac{dA}{dS} + \frac{1}{\rho}\frac{d\rho}{dS} = 0$$

or

$$\frac{dV}{V} + \frac{dA}{A} + \frac{d\rho}{\rho} = 0$$

Basically this implies that the quantities V, A and ρ cannot *all* increase, or *all* decrease, in the direction of flow.

Fig.3.2. defines the conditions appertaining to a small particle of 'ideal' fluid moving along a streamline, the cross sectional area of the particle being ΔA and its length ΔS. The weight of the particle is thus $\rho g \Delta A \Delta S$. If the pressure at the upstream end is P, and this varies continuously along the streamline, the pressure at the downstream face will be

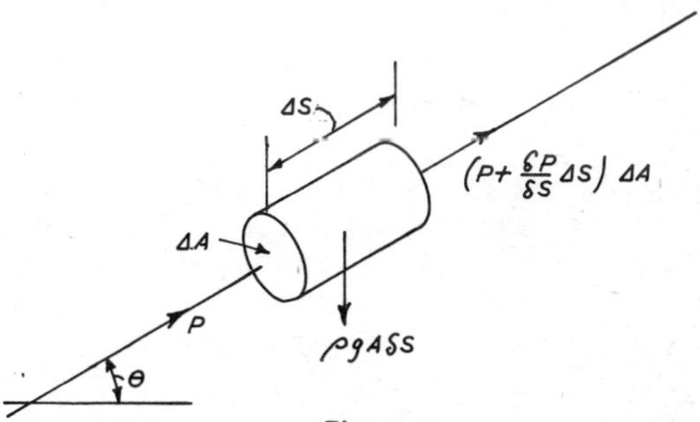

Fig. 3.2

$$\left[P + \frac{\delta P}{\delta S}\,\Delta S\right]\Delta A$$

In other words, the pressure has increased by the product of the pressure gradient and length of path considered. At the same time, if the streamline is inclined at an angle Θ with the vertical, the weight of the particle will have a component along the streamline equal to

$$\rho g\,\Delta A \Delta S \cos\Theta$$

The net force (to the right) is thus

$$P\Delta A - (P + \frac{\delta P}{\delta S}\,\Delta S)\,\Delta A - \rho g\,\Delta A \Delta S \cos\Theta$$

By Newton's second law of motion, this must be equal to the rate of increase in momentum of the particle, or specifically the product of its mass and acceleration. For steady conditions

$$\frac{dV}{dt} = \frac{dV}{dS}\,\frac{dS}{dt} = V\,\frac{dV}{dS}$$

Hence

$$-\frac{dP}{dS}\,\Delta A\Delta S - \rho g\Delta A\Delta S\cos\Theta = \frac{\rho g\Delta A\Delta S}{g}\cdot V\,\frac{dv}{ds}$$

$$\frac{1}{\rho g}\,\frac{dP}{dS} + \frac{V}{g}\cdot\frac{dv}{dS} + \cos\Theta = 0$$

or since $\cos\Theta$ can be expressed as dh/ds

$$\frac{dP}{\rho g} + \frac{VdV}{g} + dh = 0$$

This is generally known as the Euler equation for steady, one-dimensional flow with an ideal fluid. Since such a fluid is incompressible, the density is constant, and so the equation can be simplified to

$$\frac{P}{\rho g} + \frac{V^2}{2g} + h = \text{a constant}$$

In this form it is the more familiar Bernoulli equation, which may be rendered in terms of either 'pressures' or 'heads', viz:

$$P + \frac{\rho V^2}{2} + \rho gh = \text{constant total pressure}$$

where the first term is the static pressure, the second the dynamic pressure and the third the potential pressure, or:

$$\frac{P}{\rho g} + \frac{V^2}{2g} + h = \text{constant total head}$$

where the first term is the pressure head, the second the velocity head, and the third the potential head. The constant in each case is known as the *Bernoulli constant* and applies throughout the whole fluid, i.e. across streamlines as well as along streamlines.

Although the Bernoulli equations are valid only for incompressible ideal (non-viscous) fluids and steady flow they can give very close approximations to actual results with real fluids where viscous effects are low.

The classic example of the application of the Bernoilli equation is the calculation of the velocity of a fluid jet under the influence of a potential head (h) – Fig.3.3. Under static conditions, constant (total head)= O + O + h. With the jet flowing it is necessary to determine a point in the jet where the pressure is known. This is where the streamlines become parallel and the pressure is atmospheric, when the corresponding constant is

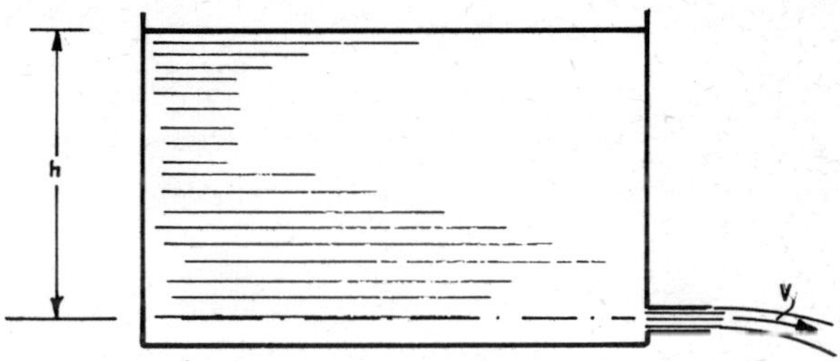

Fig. 3.3

$$O + \frac{V^2}{2g} + O$$

Thus $O + O + h = O + \dfrac{V^2}{2g} + O$

or $V = \sqrt{2gh} = 8.025\sqrt{\text{head (feet)}}$, foot per second

$= 4.44\sqrt{\text{head (metres)}}$, metres per second

(See also Table 3.1.)

It will be noticed that the jet velocity is independent of the fluid density. The solution is only exact in the absence of friction, however; and also for a nozzle form which allows the jet to develop into parallel streamlines. If a straight tapered nozzle were employed, for example, the flow would continue to contract beyond the end of the nozzle, the streamlines would be curved and the pressure would vary across the jet exit from atmospheric at the outer edge, to a value above atmospheric at the axis.

Energy Equation

Energy equations for steady flow are derived from the first law of thermodynamics which states that the amount of heat added to a fluid as it passes through a system, is equal to the change in energy content of the fluid, plus any work done by the fluid.

For a closed system:

$$q = \Delta\text{(kinetic energy)} + \Delta\text{(displacement energy)} + \Delta\text{(potential energy)}$$

For an open system:

$$q = \Delta\text{(kinetic energy)} + \Delta\text{(displacement energy)} + \Delta\text{(potential energy)} + \Delta\text{(internal energy)} + \text{(work done by fluid)}$$

The kinetic energy of any mass m is $mV^2/2$. In the case of a fluid in other than one-dimensional flow, the true kinetic energy is always greater than that calculated on the mean velocity. It can be found by integrating the product of the mass flow rate through an infinitesimal area and the kinetic energy per unit mass over the entire flow area. The ratio of the true kinetic energy per unit time to that based on the average velocity, is called the kinetic energy correction factor ($\propto$). Thus: true kinetic energy = $\frac{\propto mV^2}{2}$

For one-directional flow, $\propto = 1$. For laminar flow in a pipe $\propto = 2$; and for turbulent flow in a pipe $\propto = 1.1$ (approx.).

The potential energy is simply defined as the work required to bring a fluid to any elevation other than the datum, and is thus numerically equal to gh per unit mass, of fluid.

The internal energy is, basically, a function of the pressure and temperature of the fluid, but is more generally applicable to gases than liquids. Thermal terms are generally ignored when dealing with liquids, when an additional term is introduced to take into account the conversion of some mechanical energy into thermal energy due to viscous friction. A general expression of the energy equation then becomes

$$\frac{\propto_1 V_1^2}{2} + \frac{P_1}{\rho} + gh_1 \quad W = \frac{\propto_2 V_2^2}{2} \mp \frac{P_2}{\rho} + gh_2 + h_f^1$$

or

$$\frac{\propto_1 V_1^2}{2g} + \frac{P_1}{\omega} + h_1 - W = \frac{\propto_2 V_2^2}{2g} + \frac{P_2}{\omega} + h_2 + h_f$$

where h_f is the amount of mechanical energy converted into thermal energy, or head loss.

W = work done by, or on fluid, expressed in terms of head

ω = specific weight of fluid

In a practical system the head loss is always positive and in the direction of flow. In a hydrostatic system work is added to the system (e.g. by the pump). In the case of hydrokinetic system work is both added (by the pump) and removed (by the hydrokinetic machine).

Viscous Flow

Frictional losses arise from the fact that real fluids possess viscosity so that flow is never that of the idealised stream tube. If it were, there would be zero pressure drop along a length of pipe. Viscous losses are closely related to the formation of a boundary layer or region of retarded flow in the immediate vicinity of boundary walls, and the consequent setting up of shearing stresses in any real (viscous) fluid.

If a free fluid approaches and then flows past any plane surface, fluid particles adjacent to that surface are brought to rest, whilst those in the immediate vicinity are slowed down by the action of viscous shear – Fig.3.4. Initially, the boundary layer as formed possesses laminar flow characteristics, with a velocity gradient normal to the surface. The thickness of the boundary layer may be defined, nominally, as the depth of the region where the flow velocity is substantially retarded. The thickness of the boundary layer tends to increase downstream, which will introduce instability. As a consequence the boundary layer characteristics change from initially being laminar flow through a transistion region into turbulent flow, although the latter will still maintain a laminar sub-layer.

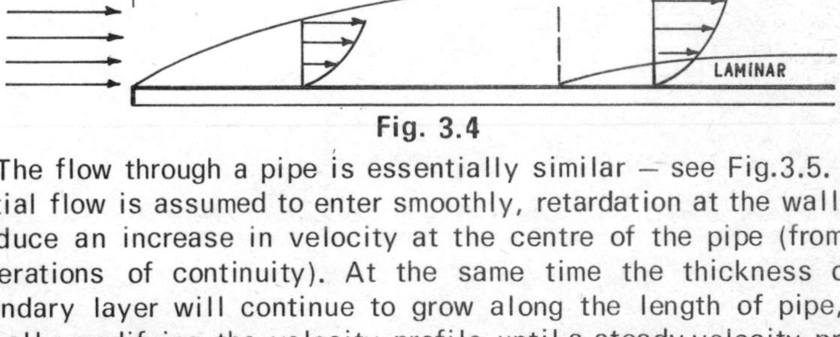

Fig. 3.4

The flow through a pipe is essentially similar – see Fig.3.5. If the initial flow is assumed to enter smoothly, retardation at the walls will produce an increase in velocity at the centre of the pipe (from considerations of continuity). At the same time the thickness of the boundary layer will continue to grow along the length of pipe, continually modifying the velocity profile until a steady velocity profile is achieved and the flow is fully developed, and will be either laminar or turbulent. In the latter case the description 'steady flow' is nominal for although a temporal mean velocity profile is formed, localised velocity differences will be apparent at points on the profile.

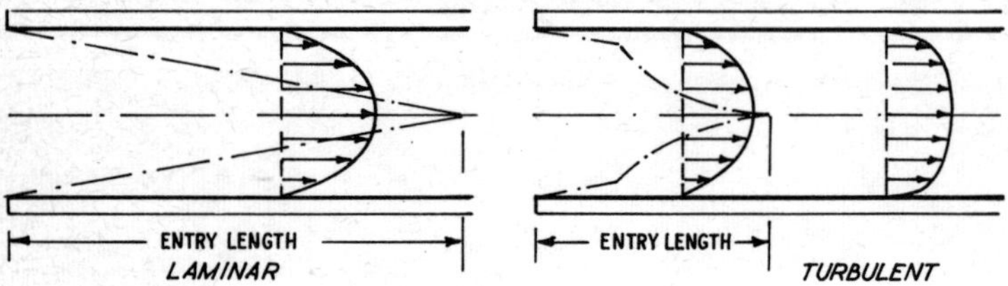

Fig. 3.5

Three parameters are involved, viz:-

(i) the shear stress in the fluid which is initially very large and then decreases in the direction of flow until it assumes a constant value with fully developed flow.

(ii) the pressure gradient which also decreases in the direction of flow until it assumes a constant value with fully developed flow.

(iii) the velocity profile changing with direction of flow until the flow is fully developed.

It is thus possible to define fully developed flow as that length yielding a constant value of one, or more, of the above parameters.

Considering an elemental length of fluid dl, by momentum requirements (Fig.3.6.).

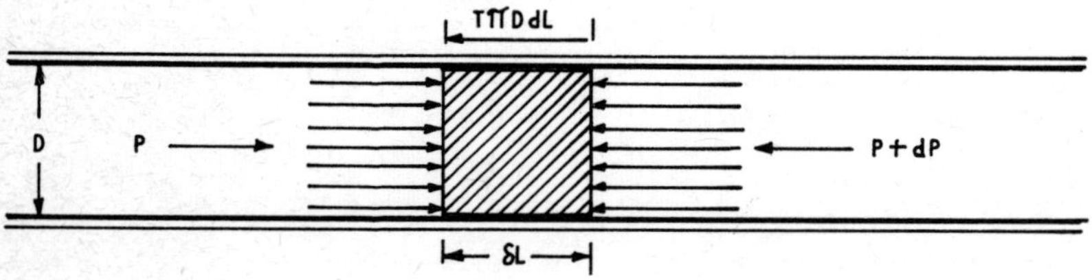

Fig. 3.6

$$P\frac{\pi D^2}{4} - (p+dp)\frac{\pi D^2}{4}] - \tau\pi D dl = \frac{\pi D^2}{4}\frac{d}{dl}(\beta\rho V^2)\, dl$$

where τ = fluid shear stress

β = momentum correction factor

This can be reduced to

$$-\frac{dp}{dl} = \frac{4T}{D} + \rho V^2 \frac{V\,d\beta}{dl} + \beta\rho V \frac{dV}{dl}$$

Before the flow is fully developed the product V1 is substantially constant (or more specifically, the quantity $\frac{V1}{\nu}$ is a constant, ν being the kinematic viscosity).

When the flow is fully developed, both $d\beta/dl$ and dV/dl are equal to zero and these terms disappear, when

$$\frac{dp}{dl} = \frac{4T}{D}$$

or, over a length L of pipe

$$\text{pressure drop } (\Delta P) - \frac{TL}{D}$$

$$\text{or } \frac{\Delta P}{L} = \frac{4T}{D}$$

The above is consistent with an incompressible fluid and Newtonian flow. The Newtonian definition of viscosity can then be applied, namely, that the shear stress is dependent on the slip velocity between adjacent layers of fluid, i.e.

$$T = \mu \frac{dV}{dD}$$

where μ is the coefficient of viscosity of the fluid.

Laminar and Turbulent Flow

Where viscous forces predominate, the fully developed flow is smooth or laminar. For a given size of pipe and a given fluid an increase in velocity will eventually bring about conditions where inertia forces become marked, modifying the flow conditions, and specifically the fully developed velocity profile. Above these critical conditions (i.e. at still higher velocities), inertia forces will predominate and the fully developed flow is entirely turbulent (except for the laminar sub-layer of the boundary layer). The ratio of the viscous to the inertia forces is thus a parameter describing the flow conditions, or type of flow which will be developed. This relationship can be expressed in terms of the Reynold's Number (RN), viz

$$RN = \frac{\rho V^2}{\mu V/D}$$

$$= \frac{\rho V D}{\mu}$$

$$= \frac{V D}{\nu}$$

where ν = kinematic viscosity (coefficient of viscosity μ divided by fluid density)

The Reynold's Number is a dimensionless quantity which can be calculated to define the type of flow in terms of flow velocity, fluid kinematic viscosity, and a characteristic linear dimension which is invariably taken as the bore diameter (D) in the case of flow through pipes.

In engineering units:

$$RN = \frac{7741.5\ VD}{\nu}\text{, or say } \frac{7750\ VD}{\nu}$$

where V is in feet per second

D is in inches

ν is in centistokes

$$RN = \frac{1000\ VD}{\nu}$$

where V is in metres/second

D is in millimeters

For calculation in terms of flow rate

$$RN = \frac{820Q}{\nu d} \text{ for Q in cu.in/sec and d in inches}$$

$$= \frac{3800Q}{\nu d} \text{ for Q in gpm and d in inches}$$

$$= \frac{21200Q}{\nu d} \text{ for Q in litres/min and d in millimetres}$$

Where the Reynold's Number does not exceed 2000, flow through pipes can be assumed to be laminar. For Reynold's Numbers from 2000 to 3000 transitional conditions apply, i.e. the flow is basically laminar but with turbulence present, although not necessarily developing into fully turbulent flow, and probably varying in extent. For Reynold's Numbers above 3000, fully turbulent flow is developed. Since the transition range represents an unstable(and varying) condition where accurate prediction is impossible, it is usual to consider all flow at Reynold's Numbers above 2000 to be turbulent.

These limits are somewhat arbitrary, but are normally adopted for general calculation. It is possible, under carefully controlled conditions, to maintain laminar flow up to Reynold's Numbers of the order of 30,000 or more. Equally, in the case of small bore pipes where the likelihood of initial disturbance is increased, the transition from laminar to turbulent flow may start at Reynold's Numbers as low as 1200. As far as practical hydraulic engineering is concerned such considerations are largely academic, since most practical flows utilised have a Reynold's Number well in excess of 2000 and are thus analysed on the basis of turbulent flow.

Pressure Drop

The pressure drop (or equivalent friction factor) for incompressible flow can be derived from first principles by applying the principle of continuity to an elemental annulus of fluid and integrating over the cross section. Since this represents only an exercise in pure mathematics, the result only will be given, viz –

$$Q = \frac{\pi D^4 \Delta P}{128 \mu L}$$

or

$$\frac{\Delta P}{L} = \frac{128 \mu Q}{\pi D^4{}_2}$$

Since

$$Q = \frac{\pi D^2}{4} v$$

Substituting for Q, we have

$$\frac{\Delta P}{L} = \frac{32 \mu v}{D^2}$$

$$= \frac{32 \rho \nu V}{D^2}$$

A more convenient form of this equation (which can also be derived by dimensional analysis) is

$$\frac{\Delta P}{L} = \frac{f}{D} \cdot \frac{\rho V^2}{2}$$

where f is a friction factor

The friction factor may also be expressed in terms of the fluid shear stress, viz –

$$T = \frac{\Delta P}{L}\left[\frac{D}{4}\right]$$

whence

$$T = f\,\frac{\rho V^2}{8}$$

Alternatively, if the friction factor is defined as the ratio of the local wall shear stress to the dynamic pressure of the mean flow –

$$T = f\,\frac{\rho V^2}{2}$$

The latter yields a friction factor of one fourth the value – i.e. 16/RN – which requires a corresponding adjustment of the basic formula.

Fig. 3.7

Laminar Flow Friction Factor

Comparison of the above equations yields the fact that for equality the friction factor must be $64\nu/VD$ or, since ν/VD is the reciprocal of the Reynold's Number,

$$f = \frac{64}{RN}$$

This holds true for all laminar flow conditions – i.e. the friction factor for use in the basic formula can be calculated directly from the Reynold's Number of the flow. It also follows that the pressure drop is independent of any pipe characteristics other than length (L) and bore diameter (D), i.e. is not affected by pipe bore condition (unless this is so rough as to modify the effective diameter, or seriously disrupt the boundary layer). Thus for laminar flow (RN below 2000) –

$$\frac{\Delta P}{L} = \left[\frac{64}{RN}\right] \frac{\rho V^2}{2D}$$

For engineering calculations it is more convenient to render this directly in terms of flow rate (Q), pipe bore (D), fluid kinematic viscosity (ν) and fluid specific gravity (SG) –

$$\frac{\Delta P}{L} = \frac{Q\nu}{14030\, D^4} \times SG$$

where ΔP is in psi
L is in feet
Q is in cu.in/sec.
D is in inches
ν is in centistokes

$$= \frac{Q\nu}{3035\, D^4} \times SG$$

where Q is in gpm (other units as above)

$$= \frac{6.52 Q\nu}{D^4} \times SG$$

where Q is in litres/min and D in millimetres

(other units as above)

$$\frac{\Delta P}{L} = \frac{28.4 Q\nu}{D^4} \times SG$$

where ΔP is in kg/cm^2

L is in metres
Q is in litres/min
D is in millimetres
ν is in centistokes

Before applying the above formulae it is necessary to check that the flow is laminar (RN less than 2000). This can be done by slide rule determination of the value of $Q/\nu D$. Flow is laminar if this value is *less* than the following critical values –

For Q in cu.in/sec and D in inches – 2.44

Q in gallons/min and D in inches – 0.53

Q in litres/min and D in millimetres – 0.094

Viscosity (ν) is in centistokes in each case.

The laminar flow velocity profile is detailed in Fig.3.8.

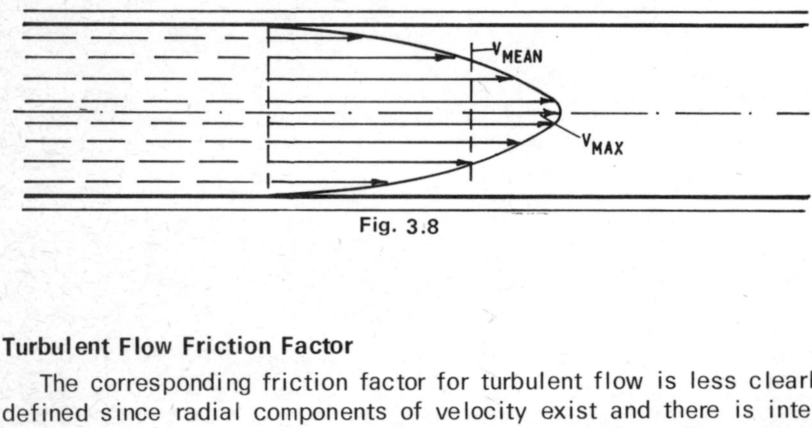

Fig. 3.8

Turbulent Flow Friction Factor

The corresponding friction factor for turbulent flow is less clearly defined since radial components of velocity exist and there is interchange of momentum between adjacent layers of fluid. The velocity profile can only be determined with reference to a temporal mean, the profile also varying with the flow Reynold's Number – Fig.3.9. Mathematical analysis is possible, but it is tedious and generally incomplete. Thus friction factors are normally determined semi-empirically for turbulent flow; and separately for varying degrees of pipe bore roughness, for this is now an influencing factor.

As far as engineering hydraulics is concerned, pipes employed are invariably smooth bore and flow is classified as 'smooth' turbulence

– i.e. surface roughness does not protrude beyond the laminary sub-layer of the boundary layer. Blasius' empirical solution for the friction factor can then be applied, which gives consistent results up to Reynold's Number 100,000, but is most accurate for Reynold's Numbers up to 10,000.

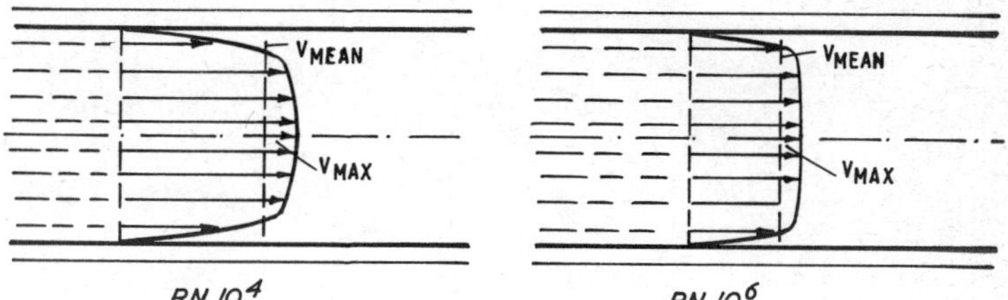

Fig. 3.9

Blasius' friction factor $f = \frac{0.3164}{RN^{.25}}$

Incorporating this friction factor in the basic equation and rendering it in its simplest form

$$\frac{\Delta P}{L} = \frac{.3164}{RN^{.25}} \quad \frac{\rho V^2}{2D}$$

$$= \frac{.3164}{8} \left[\frac{VD}{\nu}\right]^{.25} \quad \frac{4\rho V^2}{D}$$

$$= \frac{.1582\, \rho \nu^{.25}\, V^{1.75}}{D^{1.75}}$$

This equation may be used for direct calculation of pressure drop with turbulent flow in smooth bore pipes, the engineering form below being the most convenient to use –

$$\frac{\Delta P}{L} = \frac{Q^{1.75} \quad \nu^{.25}}{18750\, D^{4.75}} \times SG$$

where ΔP = pressure drop in psi
L = length of pipe in feet
Q = flow rate in cu.in/sec
D = pipe diameter in inches
ν = viscosity in centistokes
SG = specific gravity of fluid

$$\frac{\Delta P}{L} = \frac{Q^{1.75}\ \nu^{.25}}{1270\ D^4} \times SG$$

where Q is in gpm (other units as above)

Equivalent metric values are given on the pressure drop charts (see below)

In practice it is usually more convenient to employ graphical solutions – e.g. individual charts showing pressure drop per unit length for a range of pipe sizes for a given fluid viscosity. Corrections for fluid specific gravity may be applied directly by factoring.(A selection of such charts are included in the Appendix.

For more accurate solutions with high Reynold's Numbers the Nikuradse-Prandtl friction factor may be calculated. This is based on the Prandtl law of pipe friction for smooth bore pipes, verified experimentally by Nikuradse as being valid up to Reynold's Numbers of the order of 3.5 million.

Nikuradse friction factor –

$$\frac{1}{\sqrt{f}} = 2 \log (RN\sqrt{f}) - 0.8$$

Solutions for rough bore pipes are also determined semi-empirically, complete coverage normally being given graphically as in Fig.3.10. Such solutions are not normally required with hydraulic pipework.

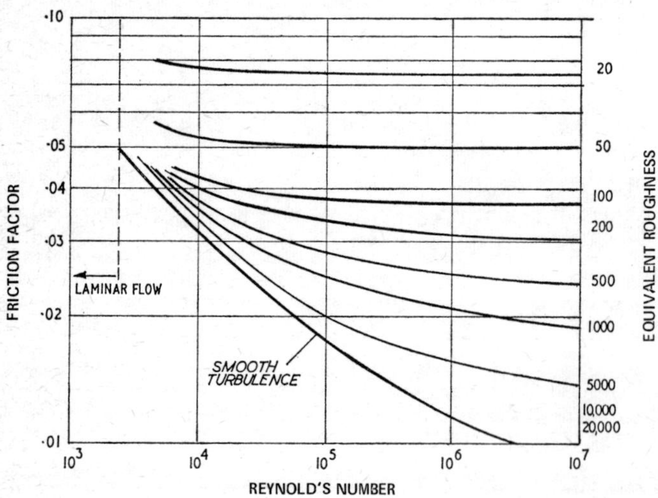

Fig. 3.10

Flow through Changing Cross-sections

Flow through changes in pipe cross section can be analysed from first principles using the Bernoulli equation (Bernoulli constant), with the subsequent introduction of an empirical coefficient to take into account frictional losses introduced with a real fluid. In problems involving closed circuits the potential head can normally be ignored in view of the inherently higher value of the other two, so that the original equation can be reduced to

$$\frac{P}{\rho} + \frac{V^2}{2g} = \text{a constant}$$

This is often more convenient to use in the form

$$V^2 + \frac{2Pg}{\rho} = \text{a constant}$$

Applying this condensed equation to steady frictionless flow along a length of horizontal pipe which contracts in cross section – Fig.3.11 the following applies –

$$V_1^{\,2} + \frac{2P_1 g}{\rho} = V_2^{\,2} + \frac{2P_2 g}{\rho}$$

i.e.

$$P_1 - P_2 = \frac{\rho}{2g}(V_2^{\,2} - V_1^{\,2})$$

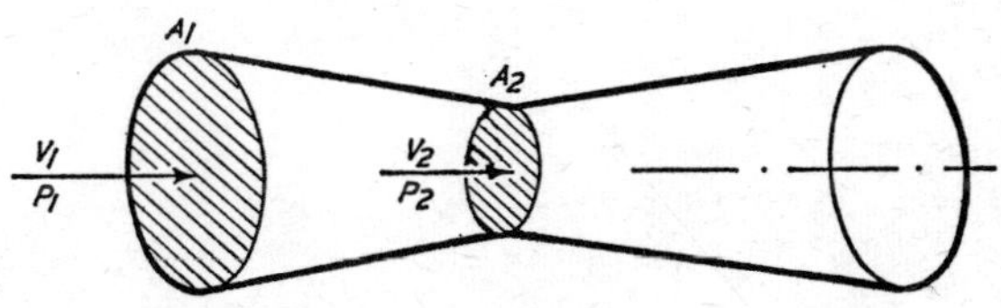

Fig. 3.11

This form of equation also shows that a steadily contracting section followed by a widening section, can be used as a principle for flow velocity measurement by measuring the difference in pressure at two extreme sections, as in the venturi – Fig.3.12. If x expresses the ratio A_2/A_1 it follows since the volume of flow is the same at both sections that $V_1 = x.V_2$ whence re-arranging the equation above and rewriting $P_1 - P_2$ as δp

$$V_1 = \sqrt{\frac{2g\delta p}{\rho} \times \frac{x^2}{1 - x^2}}$$

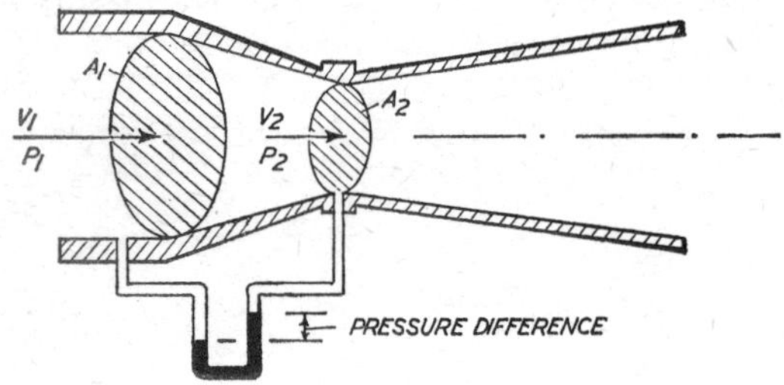

Fig. 3.12

which is the basic formula for the theoretical design of a venturi velocity meter. In practice the formula is modified by the introduction of a calibration coefficient to take into account frictional losses, etc, neglected in the Bernoulli equation.

Fundamental analysis of fluid flow through other forms of change in pipe section embody three classic cases – flow through sharp-edged orifices; sudden expansion of the flow path; and sudden contraction of the flow path. From these may be derived further solutions applicable to practical cases.

Flow through a sharp-edged orifice is particularly significant as this has a number of practical applications, and an orifice plate can also be used as an indicating flow meter (though normally restricted in application of pipes of relatively large diameter, e.g. over 2 inches).

Simple analysis is restricted to streamlined flow, the flow pattern being of the form shown in Fig.3.13. The streamlines converge on approaching the orifice and continue to converge after passing through the orifice, reaching a minimum cross sectional area, known as the *vena contracta,* downstream of the orifice before diverging again. For small circular orifices the downstream position of the *vena contracta* is of the order of half the diameter of the orifice. At the *vena contracta* all the streamlines are perpendicular to the plane of the orifice.

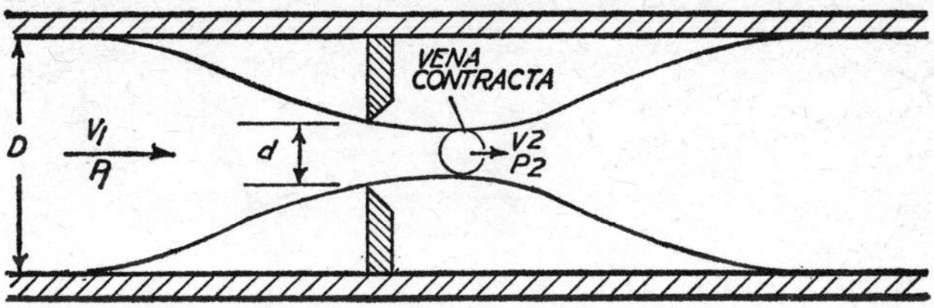

Fig. 3.13

The pressure loss, occasioned by viscous friction at the orifice, can be expressed by the Toricelli equation previously given in connection with velocity head –

$$\delta P = \frac{\rho V^2}{2}$$

This can also be derived from the Bernoulli equation, since

$$\frac{V_1^2}{2g} + \frac{P_1}{\rho} = \frac{V_2^2}{2g} + \frac{P_2}{\rho}$$

If V_1, the inlet velocity, is made zero, consistent with the original Toricelli experiment, this renders the equation in the above form. Where V_1 has a finite value, the Bernoulli equation may be re-arranged to give a solution for V_2 or the velocity at the *vena contracta* as follows –

$$\frac{V_2^2}{2g} - \frac{V_1^2}{2g} = \frac{P_1 - P_2}{\rho}$$

i.e. $$V_2^2 - V_1^2 = \frac{2g(P_1 - P_2)}{\rho}$$

But for continuity of flow $A_1V_1 = A_2V_2$

hence $$V_1 = \frac{A_2 V_2}{A_1}$$

Substituting for V_1

$$V_2^2 - \frac{A_2^2 V_2^2}{A_1^2} = \frac{2g(P_1 - P_2)}{\rho}$$

whence $$V_2^2 = \frac{2gA_1^2(P_1 - P_2)}{\rho (A_1^2 - A_2^2)}$$

$$= \frac{A_1^2}{A_1^2 - A_2^2} \left[\frac{2g}{\rho} (P_1 - P_2) \right]$$

$$\text{therefore} \quad V_2 = \frac{1}{\sqrt{1 - \frac{A_2^2}{A_1^2}}} \sqrt{\frac{2g}{\rho}(P_1 - P_2)}$$

This theoretical solution for V_2 gives a velocity which is actually less than that obtained with a real fluid since friction losses have been ignored. It can, however, be related to practical conditions by introducing an empirical factor, denoted C in this instance, which is always less than unity.

$$V_2\,(\text{practical}) = \frac{C}{\sqrt{1 - \frac{A_2^2}{A_1^2}}} \sqrt{\frac{2g}{\rho}(P_1 - P_2)}$$

Multiplying the velocity at the *vena contracta* by the cross sectional area at this point gives the flow rate viz –

$$Q = V_2 A_2 = \frac{C \,.\, A_2}{\sqrt{1 - \frac{A_2^2}{A_1^2}}} \sqrt{\frac{2g}{\rho}(P_1 - P_2)}$$

The quantity $\frac{C.\,A_2}{\sqrt{1 - \frac{A_2^2}{A_1^2}}}$ can now conveniently be expressed as a single *orifice coefficient,* leading to the following important practical formulae –

$$Q = A_2 . C_0 \sqrt{\frac{2g}{\rho}(P_1 - P_2)}, \text{ where } C_0 = \text{an orifice coefficient}$$

$$\text{pressure drop } (\Delta P) = P_1 - P_2$$

$$= \frac{\rho Q^2}{A_2 .\, C_0^2 .\, 2g}$$

Expressed in terms of the orifice diameter d, thus becomes

$$\Delta P = \frac{8\rho Q^2}{\pi^2 d^4 g C_0{}^2} \qquad \left[\text{since } A_2 = \frac{\pi d^2}{4}\right]$$

or in engineering units

$$\Delta P = 7.78 \times 10^{-5} \times \frac{Q^2}{C_0{}^2 d^4} \times \text{specific gravity of fluid}$$

where ΔP is in psi

Q = flow rate in cu in/sec

d = orifice diameter in inches

Unfortunately the orifice coefficient is not in itself a fixed quantity, being dependent both on the properties of the fluid and the geometry of the flow path so that it varies both with orifice geometry and the Reynold's Number of flow. In general, values of orifice coefficient between about .6 and .8 are considered usual and a figure of .6 or .65 is commonly adopted as typical for a sharp-edged orifice where the orifice is small compared with the tube diameter. The smaller the ratio of orifice diameter to overall (tube bore diameter) the less effect Reynold's Number has on the actual coefficient - see Fig.3.14 - which form of the curves also justifies the selection of an orifice coefficient value of the order of .6 to .65 for general solutions with small diameter orifices.

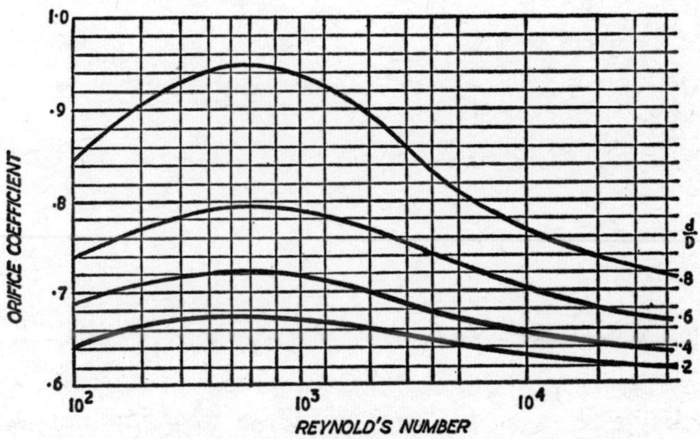

Fig. 3.14

For rapid working the engineering form of the formulae can be simplified still further by rendering the orifice coefficient in the form of a conductance coefficient embracing the fluid density, e.g.

$$\text{conductance coefficient } (Y) = \frac{\pi}{4} \cdot C_0 \sqrt{\frac{2g}{\rho}}$$

Two formulae are then available for quick calculation –

$$Q = Y\ d^2\ \sqrt{\Delta P}$$

$$\Delta P = \frac{Q^2}{Y^2 d^4}$$

In this case 'Y' is calculated for a given value of orifice coefficient (e.g. .65), applicable to a given fluid at a given temperature (i.e. employing the appropriate fluid density figure at that temperature). Alternatively, solutions may be expressed in the form of charts, either for different orifice diameters or for a specific fluid at a specific temperature - see Figs. 3.12 and 3.13.

Loss at sudden contraction

Flow behaviour at a sudden contraction follows a similar pattern to that of an orifice, a *vena contracta* being produced by the streamlines downstream from the point of contraction – see Fig.3.15.

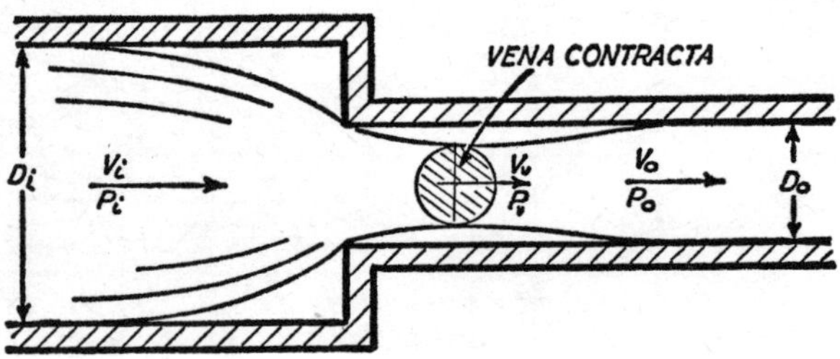

Fig. 3.15

Apart from any losses over this region there is a further loss occasioned by the enlargement of the stream cross section again to fill the pipe.

This particular case is not readily suited to theoretical analysis although numerous formulae have been developed, both theoretical and empirical. For general solutions, reasonable agreement with practice seems to be realised by treating it as an orifice case where the orifice coefficient is of the order of 1.4 and orifice diameter equal to the diameter of the smaller pipe section. A constant orifice coefficient does not, however, hold good over a range of Reynold's Numbers and it is to be anticipated that, with the equivalent orifice diameter being relatively large, considerable disagreement with practical results may be experienced at Reynold's Numbers below about 5000.

An empirical formula suggested by Keller for Reynold's Numbers greater than 5000 is –

$$\Delta P = \frac{\rho V_0^2}{2} \cdot \left[\frac{1 - \left[\frac{D_0}{D_1}\right]^2}{2}\right]$$

where the suffix 'o' refers to the outlet side and suffix 'ı' the inlet side. Other formulae of an empirical or semi-empirical nature include–

(i) $$\Delta P = \frac{\rho V_0^2}{2g}\left[\frac{1}{C} - 1\right]^2$$

where C is an appropriate orifice coefficient

$$= \frac{0.22\rho V_0^2}{g}$$

assuming an orifice coefficient value of .6.

(ii)

$$\Delta P = \frac{\rho Q^2}{4gA_0^2}$$

which is the same as an orifice loss with an orifice coefficient of $\sqrt{2}$ and diameter equal to that of the smaller pipe.

In general hydraulic work, an approximate value often taken for pressure loss is one half the value of the dynamic head in the smaller pipe.

Losses at a sudden enlargement

In the case of a sudden enlargement of pipe section, flow conditions can be analysed satisfactorily from first principles, although the energy loss (and consequent pressure drop) is largely due to the turbulence created following the change of section,Fig.3.16. Viscosity losses have only a minor effect.

The energy loss, expressed in terms of pressure loss ΔP, can be accommodated on the basis of conservation of energy, viz –

$$P_1 + \frac{\rho V_1^2}{2g} = P_2 + \frac{\rho V_2^2}{2g} + \Delta P$$

Also from considerations of momentum –

$$P_2 A_2 - P_1 A_1 = \frac{\rho A_1 V_1}{g} . V_1 - \frac{\rho A_2 V_2}{g} . V_2$$

$$= \frac{\rho A_1 V_1^2}{g} - \frac{\rho A_2 V_2^2}{g}$$

For continuity of flow –

$$\rho A_1 V_1 = \rho' A_2 V_2$$

Hence

$$\frac{P_2}{\rho} - \frac{P_1}{\rho} = \frac{V_2(V_1 - V_2)}{g}$$

Thus, by substitution

$$\frac{\Delta P}{\rho} = \frac{V_1^2 - V_2^2}{2g} - \frac{V_2(V_1 - V_2)}{g}$$

$$= \frac{(V_1 - V_2)^2}{2g}$$

Thus $\Delta P = \frac{\rho V_1^2}{2g}\left(1 - \frac{A_1}{A_2}\right)^2$, by substituting for V_2

$$= \frac{\rho V_2^2}{2g}\left(\frac{A_2}{A_1} - 1\right)^2$$, by substituting for V_1

$$= \frac{\rho}{2g}(V_1 - V_2)^2 \qquad = \frac{\rho}{2g}(V_1 - V_2)^2$$

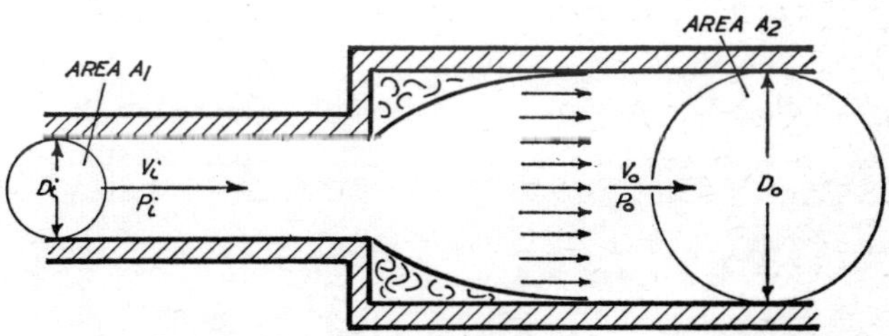

Fig. 3.16

Thus the pressure loss is dependent on the geometry only and should be independent of Reynold's Number which is generally consistent with practice over a wide range of pipe diameters, especially under turbulent flow conditions (e.g. Reynold's Number greater than 3000).

For engineering purposes it is generally easier to work with the flow rate Q rather than flow velocities, when the corresponding fomula becomes

$$\Delta P = \frac{\rho C_0 Q^2}{2gd^2}$$

where d = diameter of smaller (inlet) pipe

$$C_0 = \frac{\frac{A_0}{A_1}}{\frac{A_0}{A_1} - 1}$$

where A_0 = area of outlet pipe (A_2)
A_1 = area of inlet pipe

This formula, as a matter of interest, is consistent with that for a plain orifice, with the contraction coefficient determined by the ratio of the inlet and outlet sections, as above. The engineering form for this equation is

$$\Delta P \text{ (psi)} = 7.78 \times 10^{-5} \quad \frac{Q^2}{C_0{}^2 d^4} \times \text{S.G. of fluid}$$

where Q = flow rule in cu in/sec.

d = inlet pipe diameter in inches

Losses through annular orifices

Viscous flow through the annular space formed by two concentric cylinders is essentially similar to that between two flat, parallel plates, provided in the first instance the annular thickness, t, is small compared with the cylinder radii. The same solution will therefore apply to both cases – Fig.3.17.

There are several possible methods of analysis from first principles of which the following is based on consideration of the conditions appertaining to a small three-dimensional section of fluid sandwiched between planes of sufficient area, relative to the section thickness so that any pressure gradient can be neglected and the pressure on any plane perpendicular to the direction of flow, is uniform.

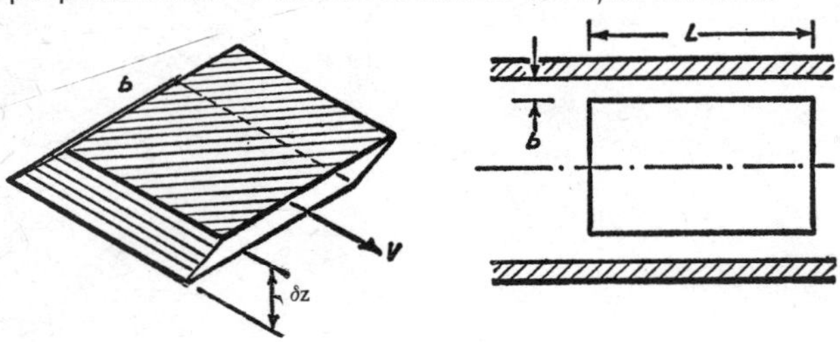

Fig. 3.17

The shear force, resultant on viscosity of the fluid and velocity on any plane or area A, due to velocity V is

$$S = \mu A \frac{dv}{dz}$$

The shear force on a plane δx distance away is then

$$S + \delta S = S + \frac{dS}{dz}\ \delta z$$

Hence

$$\delta S = \frac{dS}{dz}\ .dz$$

$$= \frac{d}{dz}\left(\mu A\ \frac{dV}{dx}\right)\delta z$$

$$= \mu A\ \frac{d^2V}{dz^2}\ .\delta z$$

Since the surface area considered is A – $b\delta x$ (the x dimension being in the plane of V)

$$\delta S = b\mu\ \frac{d^2V}{dz^2}\ dx.\delta z$$

The difference in normal pressures at each end is

$$\Delta p = bg\delta p\delta z$$

This equals the difference in shear forces

$$\Delta p = \delta S$$

$$\text{i.e.}\ bg\delta p\delta z = b\mu\ \frac{d^2V}{dz^2}\ \delta x\delta z$$

$$\text{or}\quad g\ \frac{dp}{dz} = \mu\ \frac{d^2V}{dz^2}$$

By integration

$$V = -\frac{g}{2\mu}\frac{dp}{dx}(cz - z^2)$$

where C is in the plane of Z

The flow rate, Q, is given by

$$Q = b\int_o^c V d_z$$

$$= \frac{bc^3 g}{12\mu} \cdot \frac{dp}{dx}$$

i.e. pressure drop, $\frac{dp}{dx} = \frac{12\mu Q}{bc^3 g}$

For a uniform clearance C over a distance L

$$\frac{dp}{dx} = -\frac{\Delta P}{L}$$

whence $\Delta P = \frac{12\mu LQ}{bc^3 g}$

Relating this formula for pressure drop (ΔP) to an annular orifice where the clearance space is small the total thickness of the flow path approximates to π D, where D is the diameter of the inner cylinder. The standard equation for the flow through an orifice formed by concentric, circular cylinders then becomes

$$Q = \frac{\pi D b^3 \Delta P g}{12\mu L}$$

where b is the thickness of the annulus and

L is the length of the flow path

or

$$\Delta P = \frac{12\mu LQ}{\pi D b^3 g}$$

It will be appreciated from this formula that since flow rate, or escape of fluid through a small annular orifice is proportional to the cube of the clearance space, this radial dimension is particularly critical as regards sealing – far more so than the length of the flow path provided by the annulus.

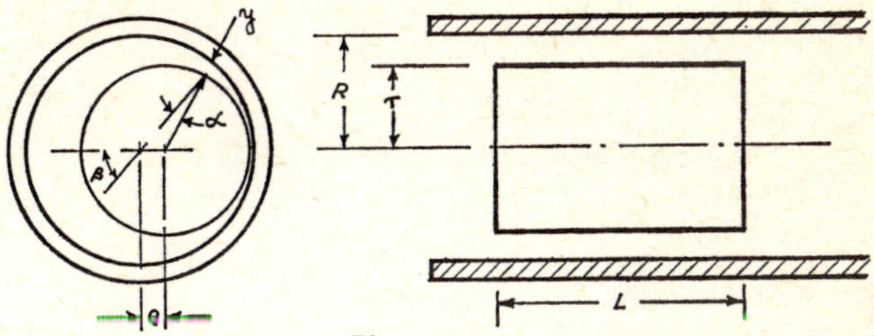

Fig. 3.18

Where the annulus is formed by two non-concentric cylinders (i.e. a piston slightly offset in the bore), the following solution has been developed by Keller (see Fig.3.18).

The piston of radius r is eccentric in the bore of radius R by the amount e.

The clearance, y, at any point $= R - (r \cos \alpha + e \cos \beta)$
which can be rewritten $y = R - (r + e \cos \beta)$

since for small clearances $\cos \alpha$ is appreciably equal to 1.

Incremental flow δq for an incremental distance $R\delta\beta$ is

$$\delta q = \frac{\Delta P y^3 R \delta\beta}{12\mu L}$$

$$= C y^3 R \delta\beta$$

where $C = \dfrac{\Delta P}{12\mu L}$

By substitution for y in the original hypothesis

$$\delta Q = C[R - (r + e \cos \beta)]^3 R\delta\beta$$

Let $R - r = b$
then

$$Q = \int_0^{2\pi} C(b - e \cos \beta)^3 R d\beta$$

and calling $\frac{e}{b}$ the relative eccentricity; designated $\bar{e}$

$$Q = \int_0^{2\pi} Cb^3 (1 - \bar{e} \cos \beta)^3 R d\beta$$

$$= 2\pi R b^3 C \left(1 + \frac{3\bar{e}^2}{2}\right)$$

$$= \frac{\pi D b^3 \Delta P}{12\mu L} \left(1 + \frac{3\bar{e}^2}{2}\right)$$

$$\text{or } \Delta P = \frac{12\mu L\, Q}{\pi D b^3 \left(1 + \frac{3\bar{e}^2}{2}\right)}$$

Heating and Cooling

Considerations of energy show that any power input which is not completely converted into mechanical work must appear as heat input in the system. This will result in a rise in temperature of the fluid, although in straightforward, low-power systems heat input and heat losses are usually balanced at an acceptable (fluid) working temperature by natural ventilation of the whole system. This applies generally where the power input is less than about 2 horsepower (1.5 kilowatts). For higher power inputs cooling may have to be introduced to maintain the fluid temperature at acceptable levels.

Typical fluid operating temperatures are summarised in Table 3.2. In general, 130 degrees F (55 degrees C) is the maximum operating temperature recommended for oil fluids for longest life, although a working temperature of 150 degrees F (66 degrees C) is usually quite acceptable, with some reduction in oil life. Thus the acceptable degree of heat input with a typical ambient temperature of 68 degrees F (20 degrees C) is that which results in a fluid temperature rise of no more than about 80 degrees F (or 35 degrees C).

The general law for the power loss in a naturally ventilated system is given by

$$\text{Power loss (HP)} = 0.00643\, K V^{0.6} S \Delta\Theta$$

where V is the airspeed in metres per second

S is the total surface area of the heat exchanger in sq metres

K is the average transfer coefficient in kcal/hour/sq.metre/°C (typically between 3 and 8)

$\Delta\Theta$ is the difference between ambient and working temperatures in degrees C

Power loss (kilowatts) = $0.0048 \, KV^{0.6} S \Delta\Theta$

where V is in metres/sec

S is in square metres

This relationship is illustrated graphically in Fig.3.19 which shows the relationship between heat exchanger area required and temperature difference, with a dissipation of the heat generated by 1 H.P. power input. The upper curve is for a compact hydraulic system. Curve 2 is for an open, well exposed system. Curve 3 is for a well ventilated system (but without forced ventilation).

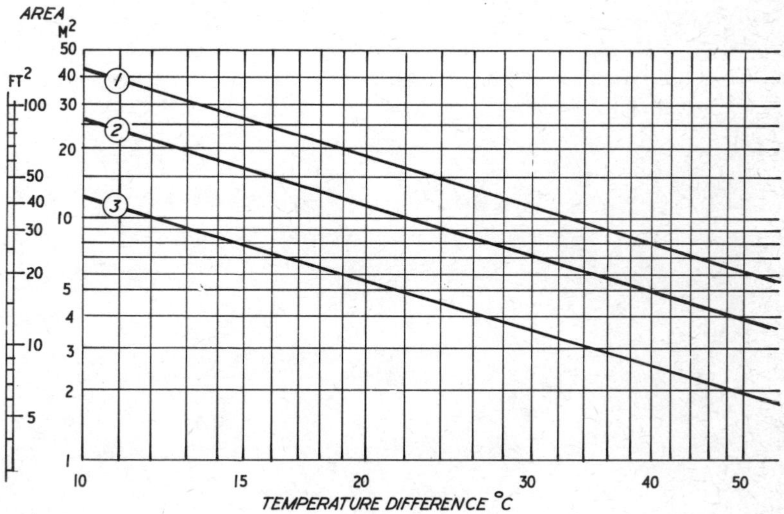

Fig. 3.19

The surface area involved is the total exposed surface area of pipework, actuators and other components, including the reservoir area. It follows that if this area is less than the figure required for the actual power loss needed to maintain the temperature difference either the system layout must be modified (i.e. to shift the heat transfer characteristics to a lower curve), or additional cooling must be applied. This can take the form of forced ventilation over a suitable surface, or intercooling via a heat exchanger.

If the difference to be met is small, the former may be the simplest and most economic solution. In this respect it should be noted that certain components, particularly in pumps and valves, the heat generation is almost invariably greater than the dissipation possibilities, and so such components are essentially fluid heaters. They should thus be located in well ventilated areas, and are a logical point at which to apply forced draught circulation of air. Additional surface area, on the other hand, is more readily achieved by increasing the length of piping (provided it is well ventilated). Although the benefit of this is partially offset by additional friction, this is a more effective method of improving 'marginal' natural cooling than increasing the size of the reservoir.

The reservoir, on the other hand, is often the component which most readily lends itself to improvement of natural cooling conditions – e.g. by locating in a position where it is in a free airstream; and in particular with the bottom of the reservoir raised so that all the surface area is exposed to the airstream.

The amount of heat transfer required (cooling) can readily be estimated by comparing the power input with the mechanical power output of the system. The difference is the power loss required, from which the total heat exchange area can be calculated directly, viz

$$S \text{ (sq metres)} = \frac{\text{HP to be dissipated}}{KV^{0.6}\ \Delta\Theta} \times 155$$

$$S \text{ (sq feet)} = \frac{\text{HP to be dissipated}}{KV^{0.6}\ \Delta\Theta} \times 510$$

This is the area which would be required with a naturally ventilated system, according to the general formula. In practice convection will (favourably) modify the actual area required, which can be determined as S/cf, where cf is the convection factor.

For a compact system, cf = 1

For an open system, cf = 2

For a well ventilated system, cf = 4

For slightly forced ventilation, cf = 6 to 8

For fan-type forced ventilation, cf = 12 to 15

Fan-type forced ventilation is the most effective method of providing efficient cooling, directed onto a radiator (similar to a car radiator) through which the fluid is circulated. This does, however, present practical problems, and also requires additional driving power. If used, it is generally better to arrange the radiator in a separate low pressure auxiliary (cooling) circuit rather than the main circuit, as this avoids the possibility of over-pressurisation.

The simpler, and more usual, solution where additional cooling has to be applied is to employ a cold water counter-flow heat exchanger, employing a nest of copper or copper alloy tubes. The average transfer coefficient (K) which can be achieved is of the order of 160 kcal/hr/m^2/°C, so the bulk of the exchanger is usually quite moderate. It is usual to install the heat-exchanger in the return circuit, but on larger systems it may be incorporated in an auxilary circuit which can be connected to or disconnected from the main circuit, as required. In some cases an external heat exchanger may be replaced by an internal coil carrying the fluid, mounted in the reservoir. Cooling is then achieved by reservoir oil flow past the coils.

Pump Heating

The pump may represent a special case, since it may be called upon to operate continuously against an intermittent demand. If the pump was left pumping its full output through a relief valve when there was no system demand, the bulk of the pump output power would be converted into heat. The resulting temperature rise would be independent of the flow rate, incidentally, and would depend primarily on the pressure energy.

$$\text{Temperature rise (C)} = \frac{\text{P (psi)} \times 0.00165}{\text{SG} \times \text{Sh}}$$

$$= \frac{\text{PC Kg (cm}^2\text{)} \times 0.0001155}{\text{SG} \times \text{Sh}}$$

where SG = specific gravity of fluid
Sh = specific heat of fluid

In practice, this condition would be avoided where the pump has to be continuously running by off-loading the pump when there is no system demand. This can be done by 'pressure off-loading' or 'flow off-loading'. In the former case the pump circulates fluid at nominal zero pressure until there is a demand from the system. In the latter case pump displacement is reduced to a nominal figure to balance internal leakage only until there is a demand from the system. In both cases the pump input horsepower is substantially reduced and only moderate heating is experienced. The actual temperature rise is given by

$$\text{Temperature rise (C)} = \frac{\text{K x pump input HP}}{\text{SG x Sh x } Q_o}$$

where Q_o = offload flow through the pump

K = 2.35 for offload flow in gpm

= 0.515 for offload flow in litres/min

Operating Temperatures

Average working temperatures for hydraulic fluids in industrial systems is of the order of 100 to 120 degrees F (38-50 degrees C). The nominal viscosity range is from about 20 to 100 centistokes, the description nominal being given to indicate that these are fluid viscosities referred to a normal ambient temperature. Ideally – and logically – the fluid viscosity is selected so as to be at an optimum value at the working temperature of the system. The fact that viscosity changes with temperature implies a number of practical difficulties.

The working temperature of any system is controllable. That is to say, the heating effect of friction (or more specifically the heat input generated when the system is working) can be offset by cooling by natural ventilation, forced cooling or intercooling, so as to avoid an excessive working temperature, or limit the fluid temperature rise to a specific maximum above the ambient. Controlled cooling can also be introduced – the simplest method being to switch an intercooler in or out of the circuit – to maintain a specific fluid temperature with changes of ambient temperature. Such precise control of temperature can be employed to give constant fluid characteristics, although it is seldom needed on the majority of systems.

Such conditions can only apply where a heat input is available, i.e. the system is working and part of the power input is converted into heat energy. Systems which work only intermittently will be subject to cooling down when idle, and if the idle period is long enough, will assume the ambient temperature. If the ambient temperature falls drastically, this will substantially increase the viscosity of the fluid, which may present difficulties in starting up. If this possibility is avoided by the use of a lower viscosity fluid, then the loss of viscosity on warming up may render the fluid characteristics unsuitable for the working of the system. Equally, in the case of a system which has to operate at a high temperature (e.g. in a higher ambient temperature), selection of a suitable viscosity for that temperature may well yield an excessive fluid viscosity on standing.

The overall problems involved can best be referred to a viscosity temperature graph on which limits can be placed. These limits are necessarily arbitrary for general solutions, but can be quite specific for individual requirements.

Dealing with upper and lower viscosity limits first, the lowest limit for viscosity with oil fluids is normally set by lubrication requirements and is of the order of 10 centistokes. A corresponding upper limit is more difficult to define, but could probably be set at

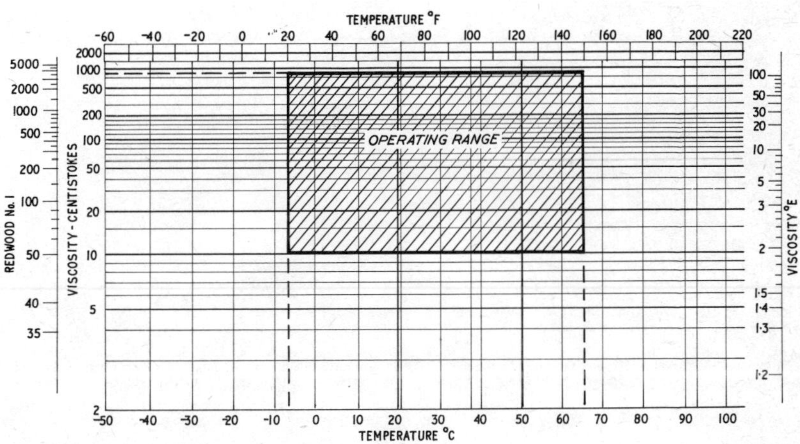

Fig. 3.20

around 800 centistokes, although a much lower figure would normally be preferred to minimise power input on starting up a cold system – see Fig.3.20.

Viscosity-temperature characteristics of individual fluids would then indicate their suitability between these limits, i.e. specifically the temperature range over which their viscosity lies between the extreme limits. Requirements are, however, likely to be more specific, particularly where the hydraulic pump is concerned. The ideal, or design, velocity in this case is set by the viscosity of the fluid at the working temperature, giving a spot point on the chart. The 'recommended' oil viscosity may well be quoted as the viscosity at normal ambient, in which case a fluid with the same viscosity index as the design fluid would have to be employed to achieve identical characteristics at the working temperature. It is thus necessary to follow closely fluid recommendations for individual pumps which will vary both with the type of pump and individual designs.

Where viscosity requirements can be defined more explicitly in terms of upper and lower limits, as in AB, CD on Fig.3.21, the usable temperature range of a particular fluid is also clearly defined, i.e. by the intersection of AB with the left hand side of the fluid curve, and CD with the right hand side of the fluid curve. Any fluid temperature between T_U and T_L will be suitable. Such limits will be specific to individual fluids, i.e. individual fluid curves.

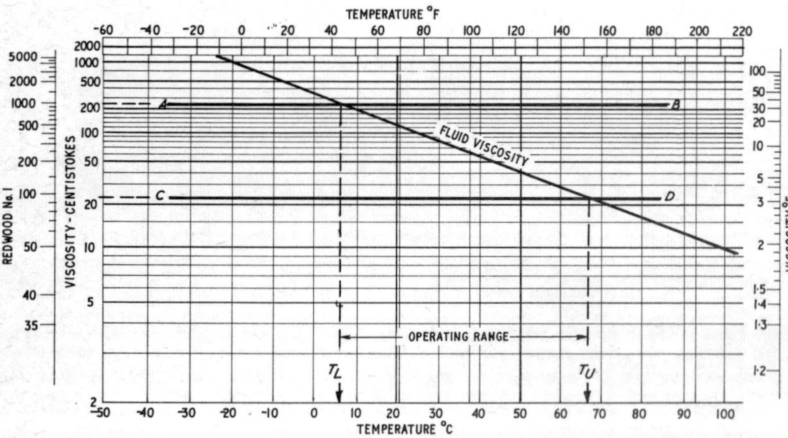

Fig. 3.21

PRESSURE LOSS SMALL BORE PIPES AND TUBES (LAMINAR FLOW)

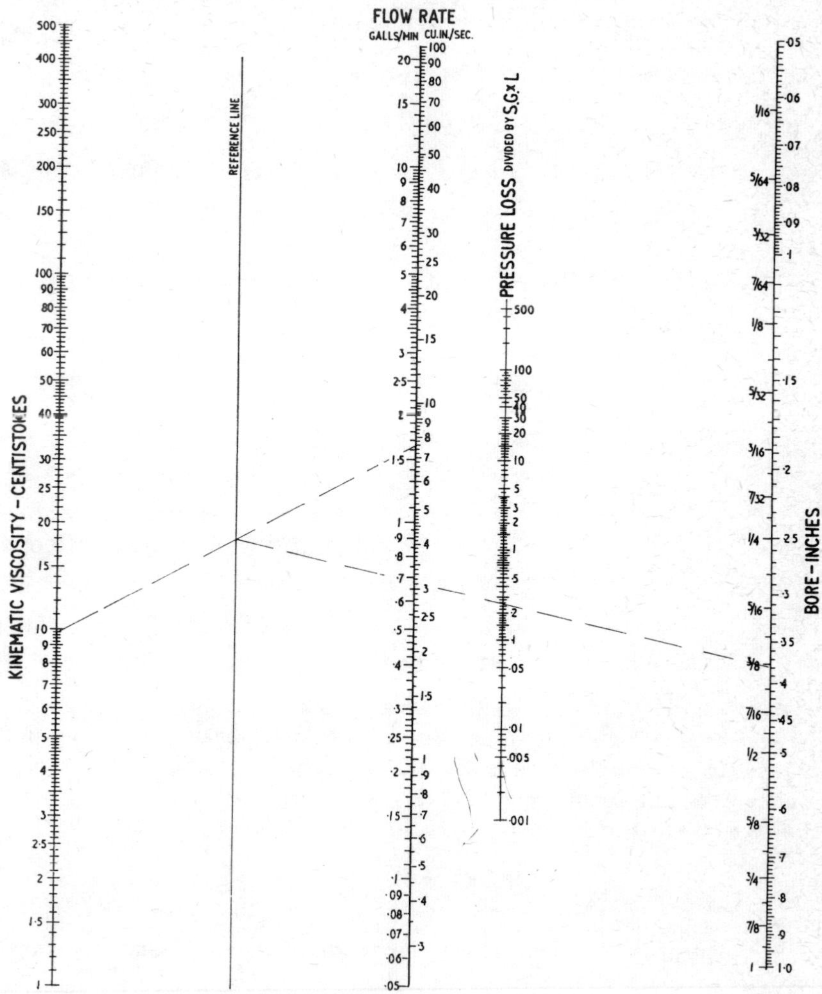

Fig. 3.22

Example : Given flow rate – 7.6 cu in sec Kinematic viscosity – 9.8 centistokes. Pipe bore – .38 ins. Connect flow rate and viscosity. From same point on reference line, connect to bore. Pressure loss read off – .25 psi. This represents pressure loss per foot run of pipe divided by S.G of fluid.

The actual working temperature range will be from the lowest ambient temperature (cold start) up to $T_U = T_a + \Delta\Theta$, where T_a is the ambient temperature and $\Delta\Theta$, the rise in temperature due to energy conversion into heat. The latter can be limited by cooling, but this does not necessarily limit the value of T_U for with a constant value for $\Delta\Theta$, T_U can shift up or down with changes in the ambient temperature T_a. However, if is $\Delta\Theta$ based on the maximum ambient temperature likely to be experienced, so that T_a (max) + $\Delta\Theta = T_U$ this will ensure that the system will never operate at a temperature above T_U.

No similar control over T_L is possible, however, since this is a natural (ambient) temperature, or $T_{a\,(min)}$. If $T_{a\,(min)}$ is lower than T_L, as indicated on the graph, then either another fluid must be chosen which provides the required maximum viscosity at this temperature, or artificial heating must be employed to make up the difference between $T_{a\,(min)}$ and T_L. An alternative choice of fluid may or may not be a practical solution, and in any case will yield a different value of T_U and possible more stringent cooling requirements. A low viscosity oil, in fact, would normally only be selected for systems which have to work continuously at very low temperatures, or intermittently at low temperatures. Fluid heating, on the other hand, is fairly straightforward and suitable for general application, and can be accomplished quite simply with an electric immersion heater in the reservoir.

The capacity of such a heater can readily be calculated from the weight of fluid in the reservoir, the specific heat of the fluid, and the rate of temperature rise. In the case of typical hydraulic oils, 1 kilowatt is required to raise 88 gallons of oil (400 litres) through 5 degrees C in one hour. The capacity for other heating requirements can be derived by simple proportion. Certain practical limitations are placed on the current density of electric heaters used in oil tanks. This should not exceed 230 watts/dm^2 for static oil (15 kw/sq in); and 300 watts/dm^2 (19.5 kw/sq in) for moving oil.

PRESSURE LOSS IN PIPES AND TUBES—TURBULENT FLOW

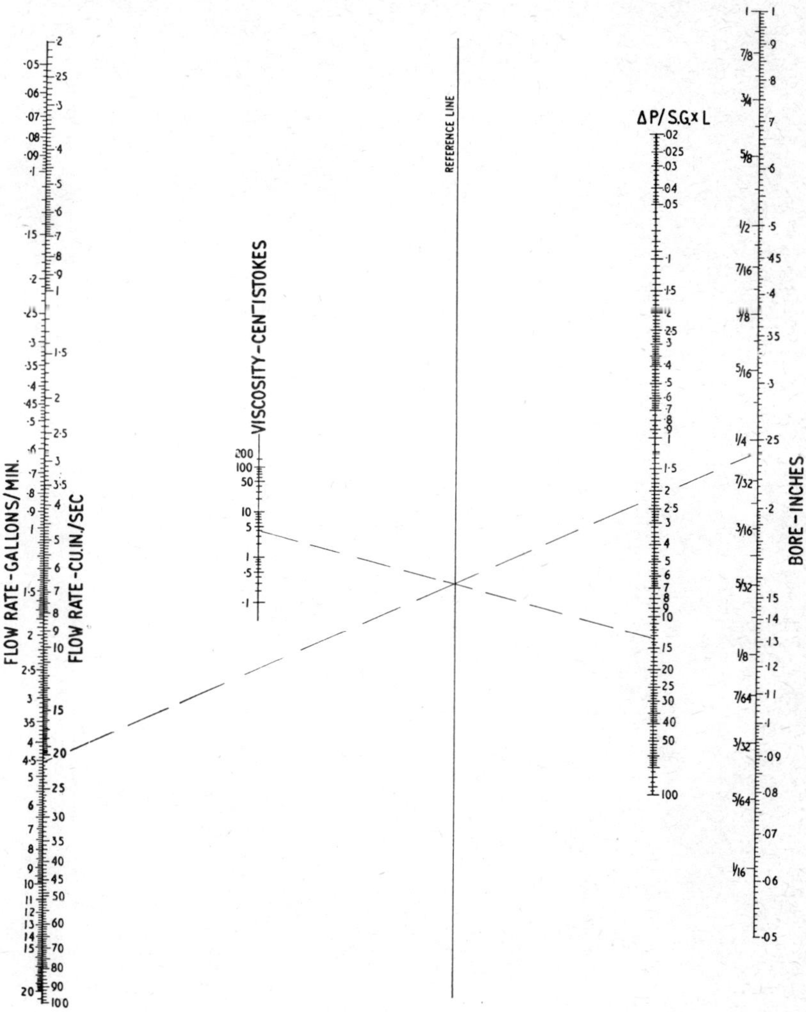

Fig. 3.23

Example : Given flow rate — 21 cu in sec : viscosity — 4 centistokes : bore — .24 ins. Connect flow rate and bore scales. From same point on reference line, join viscosity (4) to pressure loss scale. Answer : 13.5 psi. Actual pressure loss 13.5 x pipe run in feet x S.G. of fluid.

PRESSURE DROP THROUGH ORIFICE

(Fluid Specific Gravity ·849)

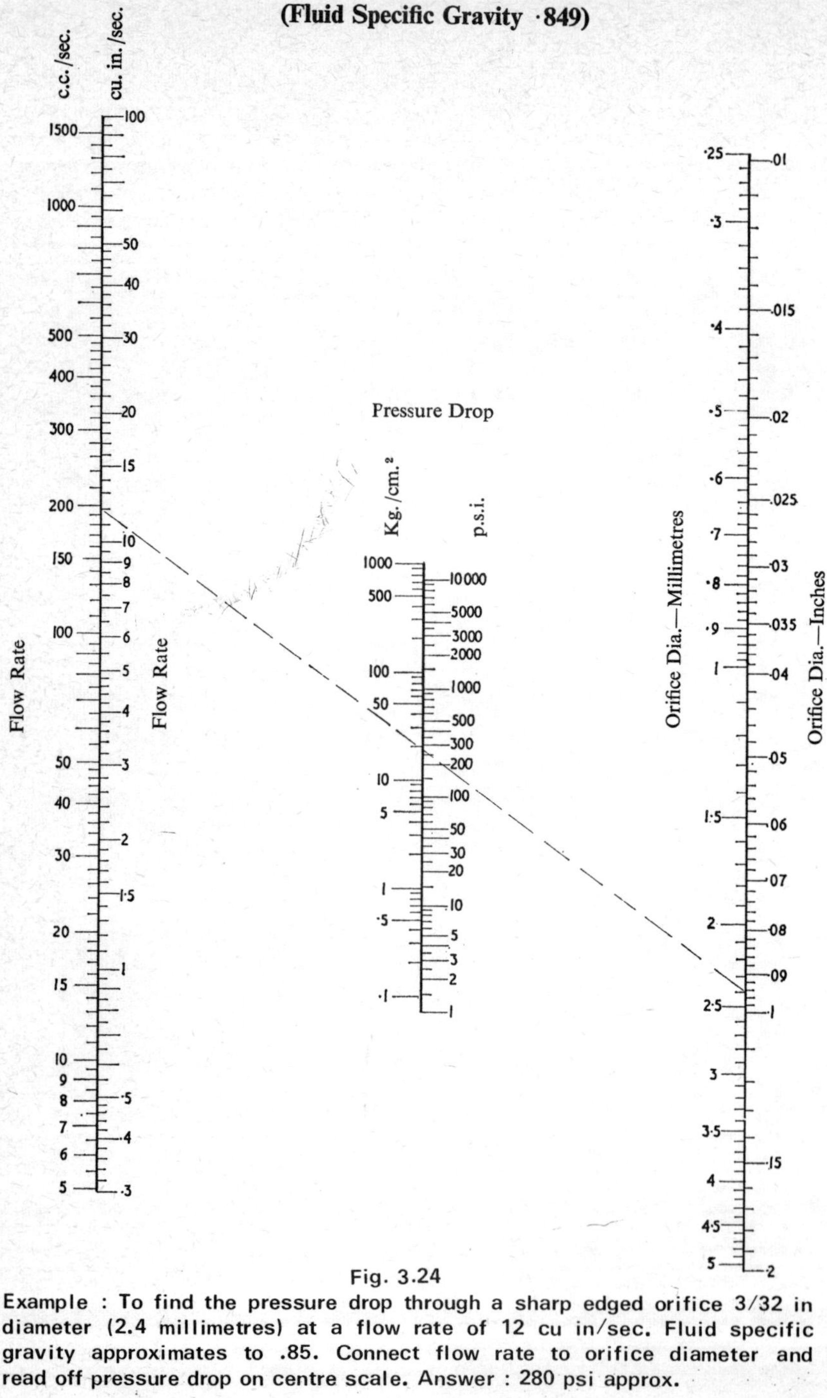

Fig. 3.24

Example : To find the pressure drop through a sharp edged orifice 3/32 in diameter (2.4 millimetres) at a flow rate of 12 cu in/sec. Fluid specific gravity approximates to .85. Connect flow rate to orifice diameter and read off pressure drop on centre scale. Answer : 280 psi approx.

TABLE 3.1
THEORETICAL FLUID JET VELOCITY
(Velocity in feet per second)

HEAD feet	0	1	2	3	4	5	6	7	8	9
0		8.025	11.35	13.90	16.05	17.94	19.66	21.33	22.79	24.03
10	25.38	26.62	27.80	29.34	30.00	31.09	32.10	33.09	34.05	34.98
20	35.89	36.77	37.64	38.48	39.31	40.13	40.92	41.70	42.40	43.21
30	43.95	44.68	45.90	46.10	46.79	47.47	48.15	48.81	49.47	50.11
40	50.75	51.39	52.01	52.62	53.23	53.83	54.43	55.02	55.60	56.18
50	56.74	57.31	57.87	58.42	58.97	59.51	60.05	60.59	61.11	61.69
60	62.16	62.68	63.19	63.69	64.20	64.70	65.19	65.69	66.18	66.66
70	67.14	67.63	68.09	68.57	69.03	69.50	69.96	70.42	70.87	71.33
80	71.68	72.23	72.67	73.11	73.55	74.00	74.42	74.85	75.25	75.71
90	76.13	76.55	76.07	77.39	77.81	78.22	78.63	79.04	79.44	79.85
100	82.25	—	—	—	—	—	—	—	—	—

TABLE 3.2
TYPICAL FLUID OPERATING TEMPERATURES

Working Conditions or System	Average Fluid Temperature °C	°F
Ideal	under 45	under 110
Satisfactory	45–55	110–130
Acceptable, with reduced oil life	55–65	130–150
Acceptable, with limited fluid life	65–140*	150–280*
Satisfactory, with intercooling	65–95	150–200

* Depending on fluid

4 LINEAR MOTION SYSTEMS

The basic hydrostatic principle of force transmission defines the force developed by a linear motion actuator as the product of the fluid pressure and the piston area over which it acts. The fluid pressure concerned is, however, that of the fluid entering the cylinder which will be lower than the pump pressure (or nominal system pressure) by virtue of the pressure drop through the interconnecting pipework – Fig 4.1.

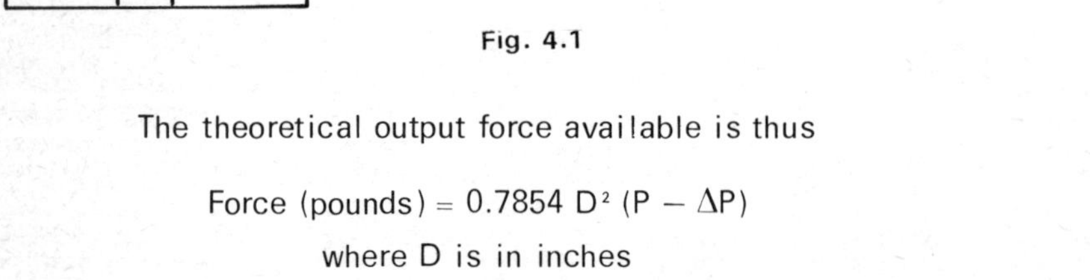

Fig. 4.1

The theoretical output force available is thus

$$\text{Force (pounds)} = 0.7854\ D^2\ (P - \Delta P)$$

where D is in inches
and P and Δ P are in psi

$$\text{Force (kilograms)} = 0.007854\ D^2\ (P - \Delta P)$$

where D is in millimetres
and P and Δ P are in kg/cm^2

Mechanical losses are also involved, due to static friction and dynamic friction. The former is generally referred to as the 'breakout' force, or the actual force which must be applied to overcome static friction and initiate movement. To a large extent this will depend on the type and condition of the piston and rod seals, and in particular whether or not the seals have dried out after a period of idleness.

In general, the breakout force will tend to be higher with smaller diameter cylinders than with larger cylinders with similar seals. The breakout force can more conveniently be expressed in terms of breakout pressure, or the minimum pressure required to overcome static friction. Typical values for a range of cylinder sizes are shown in Fig 4.2, all cylinders employing rubber-impregnated fabric seals. The higher breakout pressure required for retraction is explained by the difference in effective piston areas involved, ie a higher breakout pressure is required when the fluid is acting on an annular area only, as opposed to the full piston area.

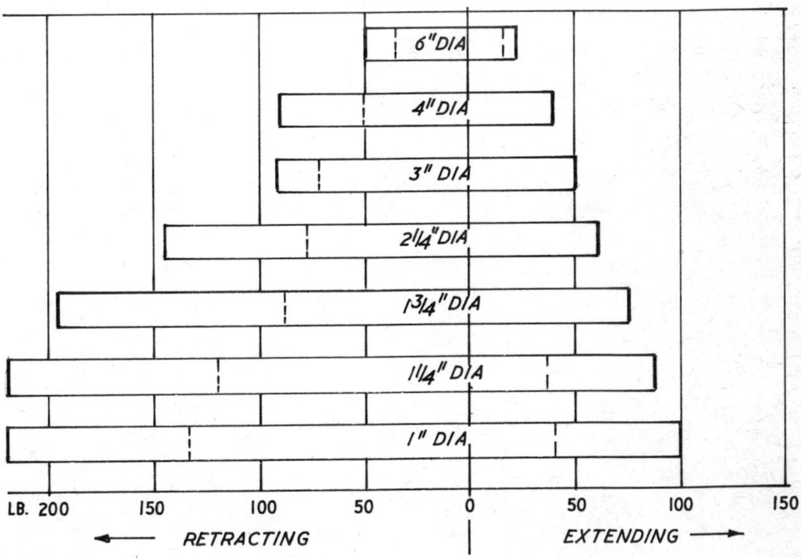

Fig. 4.2

Once motion is initiated, the frictional force opposing motion is largely a function of the seal design and the ability of the fluid to maintain optimum lubrication of the interface. Typically dynamic friction losses should never exceed 5 per cent, and should be substantially less with good seal design and adequate lubrication.

Realistically, therefore, the force available on the extension stroke should not be less than:

Force (pounds) = $0.75\ D^2\ (P - \Delta P)$, for D in inches, P in psi
or Force (kilograms) = $0.0075\ D^2\ (P - \Delta P)$, for D in mm, P in kg/cm^2

and the force available on the retraction stroke not less than:

Force (pounds) = $0.75\ (D^2 - d^2)\ (P - \Delta P)$
or Force (kilograms) = $0.0075\ (D^2 - d^2)\ (P - \Delta P)$

where d is the rod diameter, in inches or millimetres, respectively.

Some further modification of the perfomance may be produced by back pressure effects, however. In the case of a single-acting cylinder with spring return, the available hydraulic force output is reduced by the value of the spring back pressure. The return movement will itself be subject to back pressure, namely the pressure drop through the line return to the tank with the supply now switched off – Fig 4.3. For equilibrium

$$\frac{F_S}{0.75\ D^2} = \Delta P_r$$

$$\text{or } F_S = 0.75\ D^2\ \Delta P_r$$

where F_S is the spring force

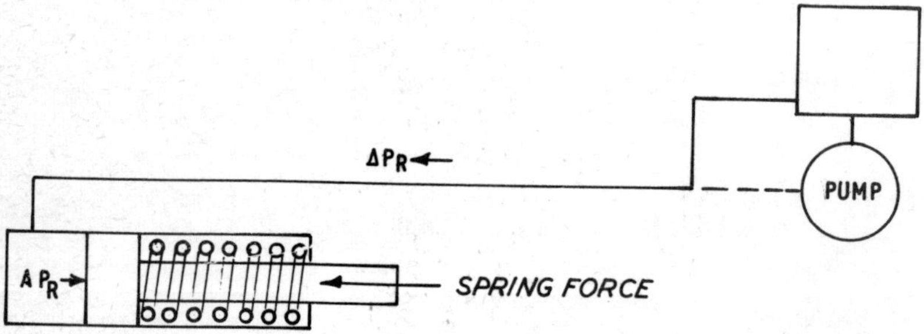

Fig. 4.3

At the same time the value of $F_S/0.7854D^2$ must be greater than the breakout pressure, otherwise return movement will not start at all. For any value of the spring force greater than this the system will be self-adjusting, ie an excess of spring force will increase the speed of return and the velocity of the return flow, which in turn will increase the pressure drop through the return line until equilibrium is reached.

The spring force required for any speed of return can be calculated on this basis, ie by analysis of the fluid motion involved.

In the case of a double-acting cylinder, both sides of the cylinder are initially charged with fluid, motion then being iniated by pressurising one side or other. It follows that the outflowing fluid on the unpressurised side of the piston must initiate a back pressure opposing motion since that fluid has to be forced along a return line with a consequent pressure drop. This pressure drop is the back pressure.

To analyse these conditions in detail it is more convenient to consider the ratio of areas involved, ie the ratio A1: (A1 – A2) equals τ, where A1 is the full piston area and A2 the rod cross sectional area. For convenience it will be assumed that the inlet and return lines are similar in size and length, and thus have equal losses for equal rates of flow; and that the speed of operation of the cylinder is to be the same in either direction.

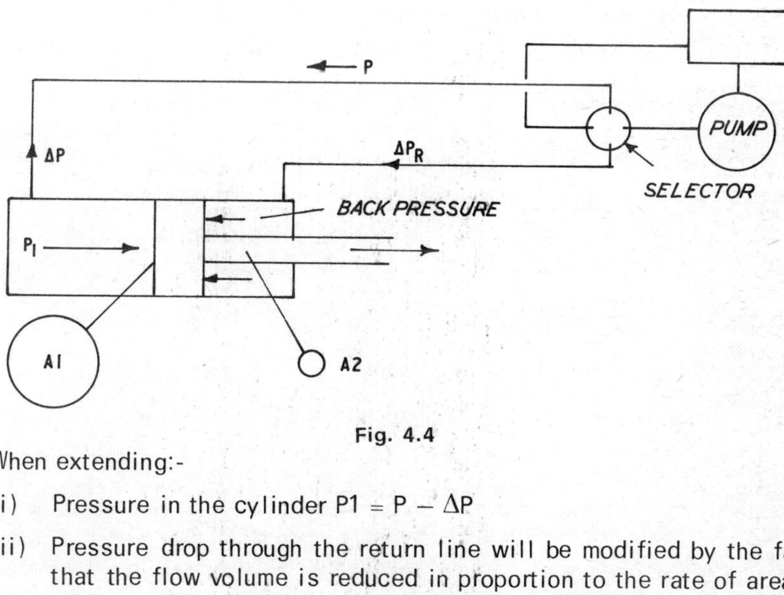

Fig. 4.4

When extending:-

(i) Pressure in the cylinder $P1 = P - \Delta P$

(ii) Pressure drop through the return line will be modified by the fact that the flow volume is reduced in proportion to the rate of areas, thus

$$\Delta P_\tau = \frac{\Delta P}{\tau} = \text{back pressure.}$$

(iii) The pressure on the other side of the piston necessary to overcome this will be

$$\frac{1}{\tau} \times \frac{\Delta P}{\tau} = \frac{\Delta P}{\tau^2}$$

(iv) Thus total loss of pressure on the inlet side will be

$$\Delta P + \frac{\Delta P}{\tau^2}$$

$$= \Delta P \left(1 + \frac{1}{\tau^2}\right)$$

When extending (Fig 4.5):

(i) Pressure in the cylinder $P_1 = P - \Delta P$

(ii) Pressure drop through return line

$$\Delta P_\tau = \Delta P/_\tau$$

(iii) Pressure on pressurised side of piston to overcome this will be ΔP_τ

(iv) Thus total pressure drop on the inlet side will be

$$\Delta P \left(\frac{1}{\tau} + \tau\right)$$

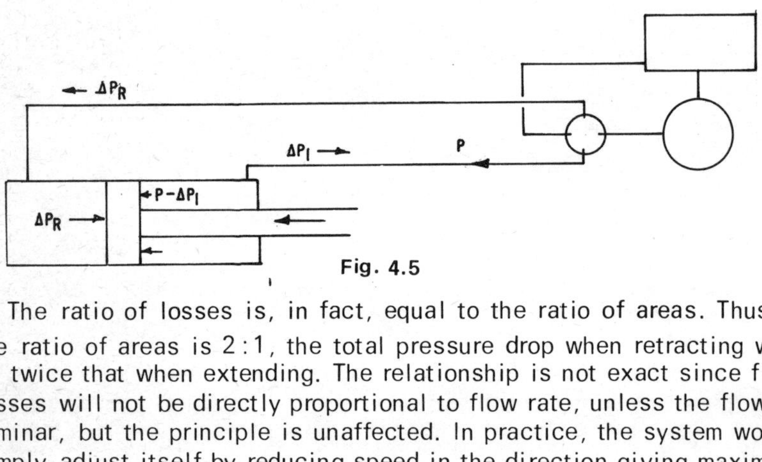

Fig. 4.5

The ratio of losses is, in fact, equal to the ratio of areas. Thus if the ratio of areas is 2:1, the total pressure drop when retracting will be twice that when extending. The relationship is not exact since flow losses will not be directly proportional to flow rate, unless the flow is laminar, but the principle is unaffected. In practice, the system would simply adjust itself by reducing speed in the direction giving maximum pressure drop. With large area ratios this reduction in speed can be considerable, and may have a significant effect on performance achieved, ie a marked reduction in the actual output force available.

In general, however, back pressure effects can normally be ignored where the area ratio is of the order of 4:3, which is commonly adopted. If it is necessary to increase the rod diameter appreciably from considerations of strength or rigidity, then back pressure effects may have to be evaluated in detail in assessing the true performance of a double-acting cylinder.

Fluid Parameters

It will also be appreciated that the pressure drop experienced is directly related to the viscosity of the fluid, favouring the use of a low viscosity fluid and also one with a high viscosity index where the cylinder may be worked over a wide range of ambient temperatures. On the other hand a relatively high viscosity is desirable from the point of view of efficient sealing, although this is seldom a critical factor with oil fluids. Satisfactory sealing is normally obtainable with conventional seals down to a fluid viscosity of about 5 centistokes. It is considerably more important that the fluid viscosity (and lubricating properties) should match the pump requirements than cylinder seal requirements. Sealing problems may be aggravated where low viscosity fluids have to be employed, however; or special fluids which have a low surface tension (eg silicone fluids). Fluids which have a surface tension of less than 30 dynes/cm are usually particularly troublesome to seal.

Internal leaks, due to poor sealing, can lead to loss of performance, and particularly loss of 'holding' ability under load.

The rigidity of the system – particularly that of a high pressure system working against heavy loads – will be directly related to the compressibility of the fluid. It should be appreciated, however, that fluid compressibility is only one of the possible causes of system 'elasticity', overall rigidity depending also on the rigidity of individual components (and particularly the mounting of actuators), and mechanical backlash which may be present. The overall *stability*, as distinct from rigidity, will depend on the sensitivity and response requirements, and the inertia of the system.

There are thus limits to both rigidity and stability with simple linear actuator systems. For particular requirements it may be better to employ an irreversible type actuator for linear force output with high rigidity (eg a screw jack driven by a hydraulic motor); or adopt a closed loop servo system circuit incorporating feedback to achieve the required stability.

Linear actuators themselves may incorporate special features to improve the rigidity of the system, and in particular to resist reactions from the load on the output member. This reaction will result in a tendency for the fluid on one side of the piston to be compressed in transmitting the reaction to the cylinder mounting. At high pressures and high loads this can lead to considerable loss of rigidity and even affect the control valve setting. One method of overcoming this is to employ a differential area piston subjected to constant fluid pressure on the annulus side, so that the position remains pressurised on both sides in a 'hold' position. Alternative methods include the use of separate pressuring pistons loaded by spring coils, mechanical spring loading, or the use of position controllers or irreversible mechanical movements.

Sizing of Cylinders

A generalised form of a typical linear motion system is designated in Fig 4.6 introducing resistors (R) in the inlet and outlet lines, representing the combined resistance to flow of pipe and components in the respective lines. Assuming that the flow will be turbulent (as it almost inevitably will in practical systems), the following relationships apply:-

$$\Delta P_1 = R_1 Q_1^{1.75} = P_S - P_1$$

$$\Delta P_2 = R_2 Q_2^{1.75} = P_2 - P_T$$

$$v = \frac{ds}{dt} = \frac{Q_1}{A_1} = \frac{Q_2}{A_2}$$

$$\eta P_1 A_1 - P_2 A_2 - F = \frac{Wf}{g} = \frac{Wd^2s}{dt^2} = \frac{W}{gA_1} \cdot \frac{dQ_1}{dt}$$

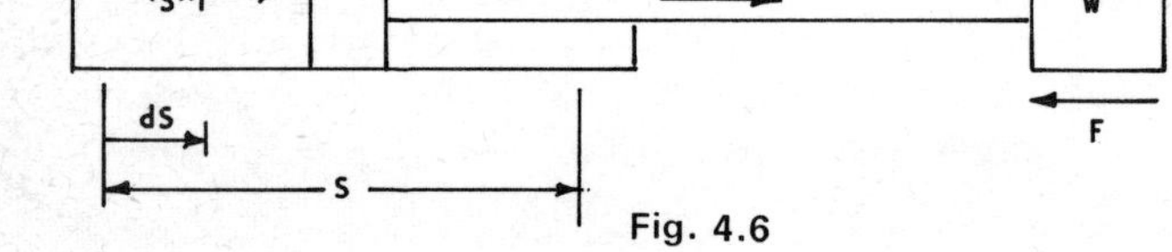

Fig. 4.6

where η is the actuator efficiency (allowing for frictional losses).

Performance under constant load and constant delivery can be determined from the appropriate formula(e).

In a majority of practical cases a much simpler solution is usually preferred. The actuator is sized on basic hydrostatic principles, ignoring back pressure effects, and 'adjusted', if necessary, by experience. Further modification of performance can then be undertaken by adjustment of the final circuit. In general, this semi-empirical approach will be satisfactory, although it does not necessarily result in an optimum size of actuator, nor does it permit of accurate performance prediction.

Such discrepancies will be even more marked if the load is varying. In such cases it is necessary to establish the kinematics of the load system for correct actuator sizing and performance prediction.

A kinematic loading curve should be determined, as in Fig 4.7. Such a curve specifies the maximum force output required from the actuator, and also the stroke required. The geometry of the cylinder can be determined on this basis, taking into account the resistance values in the circuit, and with an allowance for seal friction losses. This will give an actuator capable of performing the required duty, but not necessarily an optimum solution.

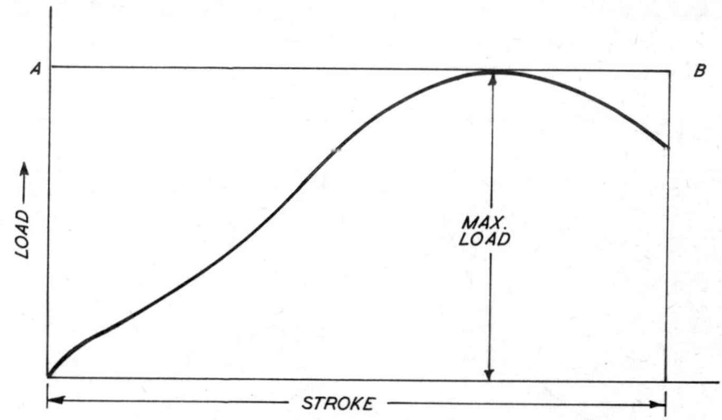

Fig. 4.7

Since the actuator is sized to meet the maximum load requirements, and can provide this output over the whole stroke, its work potential is defined by the rectangle ABCO. The efficiency of the actuator operating under the load conditions specified will then be the area under the curve relative to the area of the rectangle. The only way this can be improved is by adjustment of the kinematics of the system. In other words, the mechanical linkage should be so designed as to provide the maximum area under the curve, and thus the highest efficiency possible. Basically, this means adjusting the kinematics of the system to decrease maximum load, or stroke, or both, for the same amount of work output. This will result in a smaller cylinder being suitable for the job, because it is working more efficiently. Other basic requirements is that the ratio of the stroke to the bore should be kept to within reasonable limits (ideally between 3 and 5 to 1); and that the nearest *standard* size of cylinder should be used, regardless of the fact that this may be slightly less efficient than a 'theoretical size' cylinder. Apart from additional cost, non-standard 'theoretical' sizes of cylinders will normally not accommodate standard seal sizes, which only adds further complication and cost.

Where the load diagram cannot be adjusted satisfactorily, and is of a form which will result in very low actuator efficiency, the use of a compensator may be considered. This is an artificial load or force applied to the mechanical system modifying the resultant load in a favourable manner. Thus if the load is sharply peaked, as in Fig 4.8, the fitting of a compensator with opposite load-travel characteristics can produce a very much more favourable resultant load, as shown. Not only is the maximum load (and thus actuator size) substantially reduced, but a very high actuator efficiency can be achieved. Springs are a common form of compensator although air accumulators, acting on another cylinder, may also be used for particular applications.

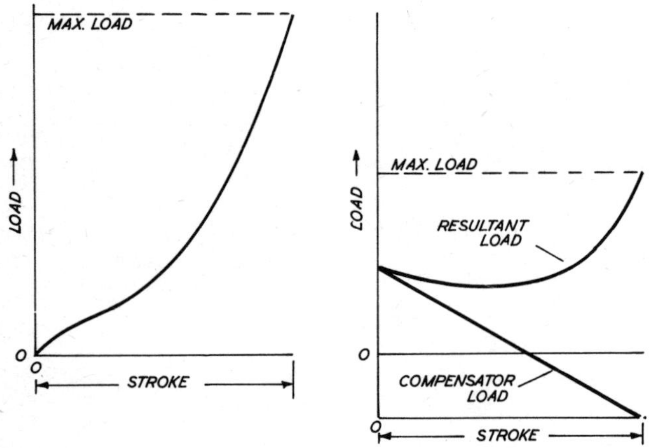

Fig. 4.8

Time of Operation

The time of operation of a cylinder is given by Q1/A1 (or Q2/A2), assuming that the pump delivery remains constant.

In the case of a variable delivery pump (or a fixed delivery pump with an accumulator in the circuit, where delivery will vary with load), the operating time is most readily determined graphically.

Two curves are required – the load-travel curve for the kinematic system; and the flow-pressure curve for the pump (or pump system) – Fig 4.9. The latter should be the true characteristics of the pump applied to the circuit, if possible.

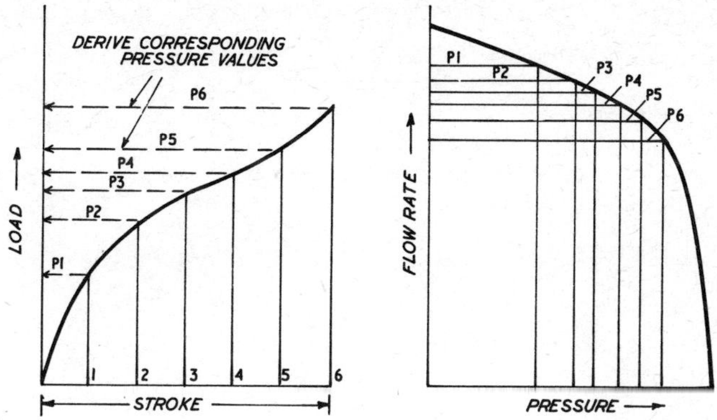

Fig. 4.9

The load-travel curve is then divided into a convenient number of points. For each point the pressure required is determined from the load divided by A1, and this figure is transferred to the pump curve to find the delivery corresponding to that pressure. Having determined the actual pump delivery at that point, the speed of movement at that point follows as Q'/A1. This is repeated for all the points and a further curve plotted of *speed* against travel – Fig 4.10.

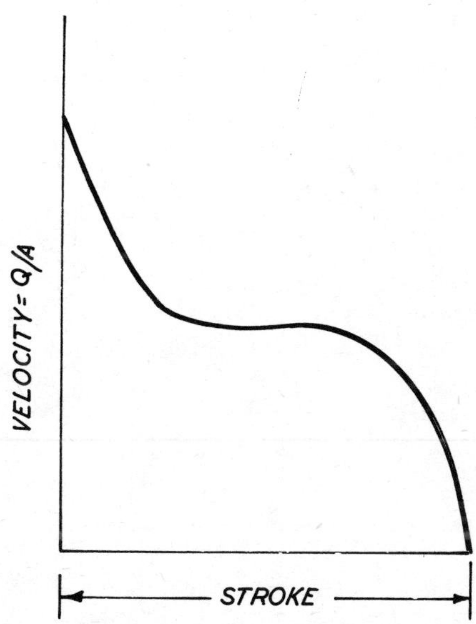

Fig. 4.10

The *time* of operation can then be determined by plotting the reciprocal of speed against travel – Fig 4.11, the time being proportional to the area under the curve.

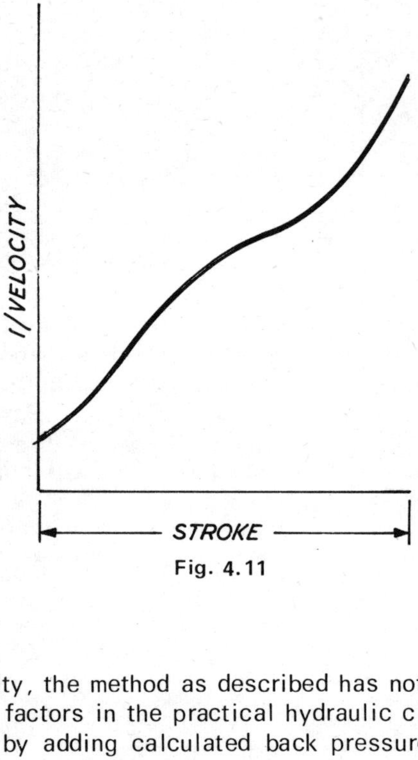

Fig. 4.11

For simplicity, the method as described has not taken into account the resistance factors in the practical hydraulic circuit. These can be accounted for by adding calculated back pressure to the load-travel curve to arrive at a true load curve, or more simply by an approximation based on the estimated time of travel determined from the simplified analysis. The time is used to calculate a mean flow rate for the system, which is added to the output load to arrive at a 'true' load curve. In either case the analysis is then reworked using the true load curve.

TABLE 4.1
THEORETICAL HYDROSTATIC OUTPUT FORCE (POUNDS)
FOR INCH SIZE CYLINDERS

Bore ins.	PRESSURE—lbs. per sq. in.								
	1,000	1,500	2,000	2,500	3,000	3,500	4,000	4,500	5,000
1/2	196	295	392	491	589	687	786	884	982
9/16	249	373	497	621	746	870	994	1,018	1,242
5/8	307	460	614	767	920	1,074	1,227	1,380	1,534
11/16	371	557	742	928	1,114	1,300	1,485	1,671	1,856
3/4	442	662	884	1,105	1,325	1,546	1,767	1,988	2,209
13/16	519	778	1,037	1,386	1,556	1,815	2,074	2,333	2,593
7/8	601	902	1,203	1,504	1,804	2,105	2,405	2,706	3,006
15/16	690	1,035	1,381	1,726	2,071	2,416	2,761	3,106	3,451
1	785	1,178	1,570	1,964	2,355	2,745	3,140	5,535	3,927
1 1/8	994	1,491	1,988	2,485	2,984	3,481	3,976	4,473	4,970
1 1/4	1,227	1,840	2,454	3,068	3,681	4,294	4,908	5,521	6,135
1 3/8	1,485	2,227	2,970	3,712	4,455	5,197	5,940	6,682	7,425
1 1/2	1,767	2,650	3,534	4,417	5,301	6,186	7,068	7,951	8,836
1 5/8	2,074	3,111	4,148	5,185	6,222	7,259	8,296	9,333	10,370
1 3/4	2,405	3,608	4,810	6,010	7,215	8,420	9,620	10,825	12,024
1 7/8	2,761	4,141	5,522	6,902	8,283	9,663	11,044	12,424	13,800
2	3,142	4,713	6,284	7,860	9,426	10,995	12,568	14,140	15,710
2 1/8	3,547	5,320	7,093	8,866	10,639	12,412	14,186	15,959	17,730
2 1/4	3,976	5,964	7,952	9,940	11,928	13,915	15,904	17,890	19,880
2 3/8	4,430	6,645	8,860	11,075	13,290	15,505	17,720	19,935	22,150
2 1/2	4,909	7,363	9,818	12,280	14,727	17,180	19,636	22,090	24,550
2 5/8	5,412	8,118	10,824	13,530	16,236	18,942	21,648	24,354	27,060
2 3/4	5,940	8,910	11,880	14,850	17,820	20,790	23,760	26,730	29,700
2 7/8	6,492	9,738	12,984	16,230	19,476	22,722	25,968	29,214	32,460
3	7,069	10,603	14,138	17,680	21,207	24,740	28,276	31,810	35,350
3 1/8	7,670	11,505	15,340	19,175	23,010	26,845	30,680	34,515	38,350
3 1/4	8,296	12,444	16,592	20,740	24,888	29,035	33,184	37,330	41,480
3 3/8	8,946	13,519	17,892	22,465	26,838	31,411	35,784	40,357	45,730
3 1/2	9,621	14,431	19,242	24,060	28,863	33,675	38,484	43,295	48,110
3 5/8	10,321	15,481	20,642	25,802	30,963	36,123	41,284	46,444	51,600
3 3/4	11,045	16,567	22,090	27,620	33,135	38,660	44,180	49,705	55,230
3 7/8	11,793	17,689	23,586	29,482	35,379	41,275	47,172	53,068	58,960
4	12,566	18,849	25,132	31,420	37,698	43,980	50,264	56,545	62,830
4 1/8	13,364	20,046	26,728	33,410	40,092	46,774	53,456	60,138	66,820
4 1/4	14,186	21,279	28,372	35,470	42,558	49,659	56,744	63,835	70,930
4 3/8	15,033	22,549	30,066	37,582	45,099	52,615	60,132	67,648	75,160
4 1/2	15,904	23,856	31,808	39,760	47,712	55,665	63,616	71,570	79,520
4 5/8	16,800	25,200	33,600	42,000	50,400	58,800	67,200	75,600	84,000
4 3/4	17,721	26,581	35,442	44,310	53,163	62,025	70,884	79,745	88,610
4 7/8	18,665	27,997	37,330	46,662	55,995	65,327	74,660	83,992	93,320
5	19,635	29,447	39,270	49,090	58,895	68,725	78,540	88,345	98,180

TABLE 4.1

THEORETICAL HYDROSTATIC OUTPUT FORCE (Pounds) FOR INCH SIZE CYLINDERS

Bore ins.	PRESSURE—lbs. per sq. in.								
	1,000	1,500	2,000	2,500	3,000	3,500	4,000	4,500	5,000
5 1/8	20,629	30,943	41,258	51,572	61,887	72,201	82,516	92,830	103,140
5 1/4	21,648	32,472	43,296	54,120	64,944	75,770	86,592	97,365	108,240
5 3/8	22,691	34,036	45,382	56,727	68,073	79,418	90,764	102,010	113,450
5 1/2	23,758	35,637	47,516	59,400	71,274	83,155	95,032	106,910	118,790
5 5/8	24,850	37,275	49,700	62,125	74,550	86,975	99,400	111,825	124,425
5 3/4	25,967	38,950	51,934	64,920	77,901	90,885	103,865	116,850	129,840
5 7/8	27,109	40,163	54,218	67,272	81,327	94,381	108,520	121,570	130,540
6	28,274	42,411	56,548	70,690	84,822	98,960	113,000	127,235	141,370
6 1/8	29,465	44,197	58,930	73,662	88,395	103,130	117,860	132,590	147,325
6 1/4	30,680	46,020	61,360	76,700	92,040	107,380	122,720	138,060	153,400
6 3/8	31,919	47,879	63,838	79,798	95,757	111,720	127,680	143,640	159,595
6 1/2	33,183	49,770	66,366	82,960	99,549	116,140	132,730	149,325	165,920
6 5/8	34,472	51,608	68,944	86,180	103,420	120,650	137,890	155,120	172,360
6 3/4	35,785	53,680	71,570	89,470	107,355	125,250	143,140	161,040	178,930
6 7/8	37,122	55,683	74,244	92,805	111,370	129,930	148,490	167,050	185,610
7	38,485	57,730	76,970	96,220	115,460	134,700	153,940	173,190	192,430
7 1/8	39,871	59,807	79,742	99,678	119,610	139,550	159,480	179,420	199,355
7 1/4	41,282	61,920	82,564	103,210	123,850	144,485	165,130	185,775	206,410
7 3/8	42,718	64,077	85,436	106,795	128,150	149,510	170,870	192,230	213,590
7 1/2	44,179	66,270	88,358	110,450	132,540	154,625	176,720	198,810	220,090
7 5/8	45,664	68,496	91,328	114,160	136,990	159,820	182,660	205,490	228,320
7 3/4	47,173	70,760	94,346	117,940	141,520	165,105	188,690	212,280	235,870
7 7/8	48,707	73,060	97,414	121,770	146,120	170,470	194,820	219,180	243,535
8	50,265	75,400	100,530	125,670	150,800	175,930	201,060	226,200	251,330
8 1/8	51,849	77,773	103,700	129,620	155,550	181,470	207,400	233,320	259,245
8 1/4	53,456	80,180	106,910	134,370	160,370	187,095	213,820	240,555	268,730
8 3/8	55,088	82,632	110,180	137,720	165,260	192,810	220,350	247,900	275,440
8 1/2	56,745	85,120	113,490	141,870	170,240	198,610	226,980	255,360	283,730
8 5/8	58,426	87,640	116,850	146,065	175,280	204,490	233,700	262,920	292,130
8 3/4	60,132	90,200	120,260	150,330	180,040	210,460	240,520	270,600	300,600
8 7/8	61,862	92,792	123,720	154,655	185,590				
9	63,617	95,420	127,230	159,050	190,850				
9 1/8	65,397	98,095	130,790	163,490	196,190				
9 1/4	67,201	100,800	134,400	168,010	201,600				
9 3/8	69,029	103,540	138,060	172,570	207,090				
9 1/2	70,882	106,320	141,760	177,210	212,650				
9 5/8	72,760	109,140	145,520	181,900	218,280				
9 3/4	74,662	111,990	149,320	186,660	223,990				
9 7/8	76,589	114,880	153,170	191,470					
10	78,540	117,810	157,080	196,350					

TABLE 4.2
THEORETICAL OPERATING TIME OF INCH SIZE CYLINDERS
(Neglecting Back Pressure Effects)

Bore ins.	TRUE DELIVERY GPM									
	1	2	3	4	5	6	7	8	9	10
1	.17	.09	.06	,04	,03	,03	,02	,02	.02	.02
1¼	.27	.13	.09	.07	.05	.04	.04	.04	.03	.03
1½	.38	.19	.13	.10	.08	.06	.06	.05	.04	.04
1¾	.52	.26	.17	.13	.10	.09	.07	.07	.06	.05
2	.68	.34	.23	.17	.14	.11	.10	.09	.08	.07
2¼	.86	.43	.29	.22	.17	.14	.12	.11	.10	.09
2½	1.06	.53	.35	.27	.21	.18	.15	.13	.12	.11
2¾	1.28	.64	.43	.32	.26	.21	.18	.16	.14	.13
3	1.53	.77	.51	.38	.31	.26	.22	.19	.17	.15
3¼	1.78	.89	.59	.45	.36	.30	.25	.22	.20	.18
3½	2.08	1.04	.69	.52	.42	.35	.30	.26	.23	.21
3¾	2.39	1.20	.80	.60	.48	.40	.34	.30	.27	.24
4	2.72	1.36	.91	.68	.55	.45	.39	.34	.30	.27
4¼	3.06	1.53	1.02	.77	.61	.51	.44	.39	.34	.31
4½	3.44	1.72	1.15	.86	.69	.57	.49	.43	.38	.34
4¾	3.84	1.92	1.28	.96	.77	.64	.55	.48	.43	.38
5	4.24	2.12	1.41	1.06	.85	.71	.61	.53	.47	.42
5¼	4.68	2.34	1.56	1.17	.94	.78	.67	.59	.52	.47
5½	5.13	2.67	1.71	1.33	1.03	.86	.73	.67	.57	.51
5¾	5.60	2.80	1.86	1.40	1.12	.93	.80	.70	.63	.56
6	6.1	3.1	2.0	1.6	1.2	1.0	.87	.76	.68	.61
7	8.3	4.2	2.9	2.1	1.7	1.3	1.2	1.0	.93	.83
8	10.9	5.5	3.6	2.7	2.2	1.8	1.6	1.4	1.2	1.1
10	17.0	8.5	5.7	4.3	3.4	2.8	2.2	2.1	1.9	1.7

NOTE: time is given in seconds per inch of stroke movement. To find time for a given stroke, multiply by stroke in inches.

TABLE 4.3
THEORETICAL HYDROSTATIC OUTPUT FORCE FOR METRIC SIZE CYLINDERS

Bore mm	Force* Kg	Bore mm	Force* Kg	Bore mm	Force* Kg	Bore mm	Force* Kg
1	.785	26	530.93	51	2042.8	76	4536.5
2	3.142	27	572.56	52	2123.7	77	4656.6
3	7.069	28	615.75	53	2206.2	78	4778.4
4	12.57	29	660.52	54	2290.2	79	4901.7
5	19.64	30	706.86	55	2375.8	80	5026.6
6	28.27	31	754.77	56	2463.0	81	5153.0
7	38.48	32	804.25	57	2551.8	82	5281.0
8	50.27	33	855.30	58	2642.1	83	5410.6
9	63.62	34	907.92	59	2734.0	84	5541.8
10	78.54	35	962.11	60	2827.4	85	5674.5
11	95.03	36	1017.9	61	2922.5	86	5808.8
12	113.10	37	1075.2	62	3019.1	87	5944.7
13	132.73	38	1134.1	63	3117.3	88	6082.1
14	153.94	39	1194.6	64	3217.0	89	6221.1
15	176.72	40	1256.6	65	3318.3	90	6361.7
16	201.06	41	1320.3	66	3421.2	91	6503.9
17	226.98	42	1385.5	67	3525.7	92	6647.6
18	254.47	43	1452.2	68	3631.7	93	6792.9
19	283.53	44	1520.5	69	3739.3	94	6939.8
20	314.16	45	1590.4	70	3848.5	95	7088.2
21	346.36	46	1661.9	71	3959.2	96	7238.3
22	380.13	47	1735.0	72	4071.5	97	7389.8
23	415.48	48	1809.6	73	4185.4	98	7543.0
24	452.39	49	1885.8	74	4300.9	99	7697.7
25	490.87	50	1963.5	75	4417.9	100	7854.0

*The theoretical output force, in kilograms, is given for a fluid pressure of 100 kg/cm^2. To find the output force for any other pressure, multiply by actual pressure in kg/cm^2 and divide by 100

5 ROTARY MOTION SYSTEMS

Three types of machines can be considered under this heading.

(i) Rotary actuators, which may be further sub-divided into (a) semi-rotary types and (b) multi-revolution types.

(ii) Hydrostatic transmissions.

(iii) Hydrokinetic devices.

The rotary output from semi-rotary actuators is derived either from piston or vane movement under fluid pressure. Basic hydrostatic principles apply as regards determination of performance, ie they are essentially hydrostatic machines. Vane types will generally provide the maximum rotational movement, eg from 220 to 270 degrees with a single vane, but less than 180 degrees with a double vane. The advantage of the latter configuration is that torque and power are doubled since fluid pressure is applied simultaneously to two vanes connected to a common spindle.

With a piston-type semi-rotary actuator, hydraulic performance can be analysed in the same way as for cylinders. The overall efficiency will be governed largely by the kinematics of the output linkage to the load and will be at a maximum when the load-travel envelope is rectangular (see Chapter 4).

With a vane movement sealing problems are aggravated since the sealing length is considerably longer than that of a piston of comparable displacement. Special tip seals are normally fitted, but some internal leak normally still remains. Vane-type actuators are thus less suited to 'holding' duties than piston-type actuators. Internal leak can be minimised by:-

(i) employing a high viscosity fluid;

(ii) applying cooling to maintain the fluid at a low temperature, if the working conditions tend to produce marked fluid heating;

(iii) offsetting fluid leakage by make-up fluid from a separate supply.

Sealing difficulties also normally limit maximum working pressure to about 1 000 psi (70 kg/cm^2). The torque obtainable is given by

$$\text{Torque} = \frac{L\,P\,(D^2 - d^2)}{N} \times \eta$$

where L = length of vane

D = diameteroverall vane (internal diameter of casing)

d – spindle diameter

P = inlet fluid pressure

N = 8 for a single vane; or 4 for a double vane

η = pressure energy conversion efficiency (allowing for internal leakage and seal friction), which may range from about 0.75 up to a maximum of about 0.95

Multi-revolution rotary actuators are normally of piston type, the piston movement driving a recirculating ball and thread movement, or similar device. This is essentially a hydrostatic device with low overall efficiency due to the low mechanical efficiency; and hydrokinetic devices or hydraulic motors are normally preferred for such duties.

Hydrostatic Transmissions

Hydrostatic transmissions comprise, basically, a hydraulic pump driving a hydraulic motor. These may be coupled in the following alternative combinations:

(i) a constant delivery pump and a constant capacity motor

(ii) a constant delivery pump and a variable capacity motor

(iii) a variable delivery pump and a constant capacity motor

(iv) a variable delivery pump and a variable capacity motor

(v) shunt coupling

(vi) differential coupling.

Only the dynamics of such systems differ. As far as the hydraulics are concerned, and the fluid requirements, performance is determined on the basis of the hydrodynamic principles involved. These define the relationship between the pump speed and motor speed, the flow through the circuit, the displacement of the machines, and the pressure in the circuit. The fluid itself is essentially 'passive' in character, whereby the pressure energy of the fluid is utilised with no significant change in fluid velocity.

The following basic equations then apply:-

(1) Power output of pump = $\eta_p \, HP_i$, where η_p = pump efficiency

(2) Power output to motor = $\eta_p \, HP_i - f_p$,
where f_p = friction power losses in connecting lines

(3) Power output of motor = $\eta_m(\eta_p \, HP_i - f_p)$,
where η_m = motor efficiency

(4) Hydraulic power = $\dfrac{QP}{k_p}$ where k_p = horsepower constant for units employed

(5) Pump output (Q) = $\dfrac{qN}{k_v}$ where q = output per revolution
N = speed of pump, rpm
k_v = volume constant for units of capacity employed

(6) Torque $T = \dfrac{qP}{2\pi}$

(7) Motor speed, $N = \dfrac{k_v Q}{q}$

It follows from equation (6) that for a constant displacement system and constant pressure the torque is also constant and independent of speed. The speed at which the motor turns is governed by the q of the motor and the actual delivery of the pump – equation (7). If pump and motor have the same displacement per revolution, then pump and motor will run at the same speed.

The behaviour of a variable displacement system follows similarly, only in this case either Q (pump) or q (motor), or both, are variable. It will be appreciated that this will give an infinitely variable stepless speed range over the extreme limits of Q (pump) and/or q (motor).

In the case of a constant displacement system the fluid velocity, and thus fluid line losses, is constant, as well as torque and speed.

With a variable-displacement system, output torque, output speed and power input are all variable and characteristic of the particular coupling employed – eg see Fig 5.1. Such diagrams define the operating range of the system, but may require to be modified to take into account power losses not included in the original computation.

Fig. 5.1 HYDROSTATIC TRANSMISSIONS.
1. 5,000 lb/in2 constant pressure line.
2. Motor stroke. 3. Pump stroke.
4. 15 h.p. line.

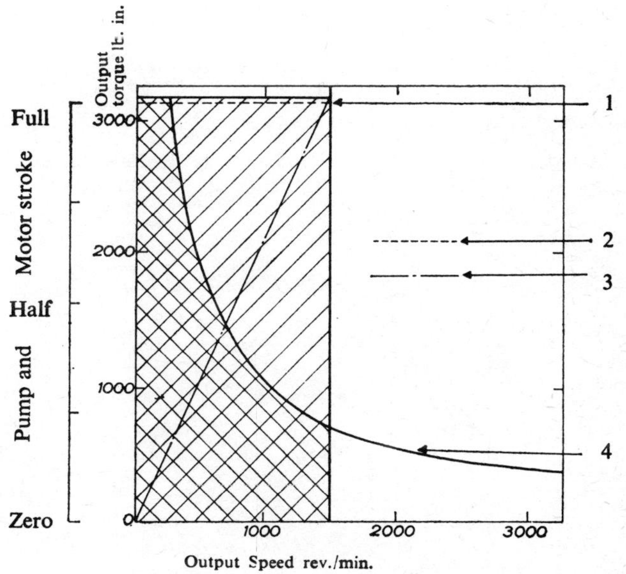

(A) Variable delivery pump. Fixed displacement rotor.

Fig. 5.1 HYDROSTATIC TRANSMISSIONS CONT.

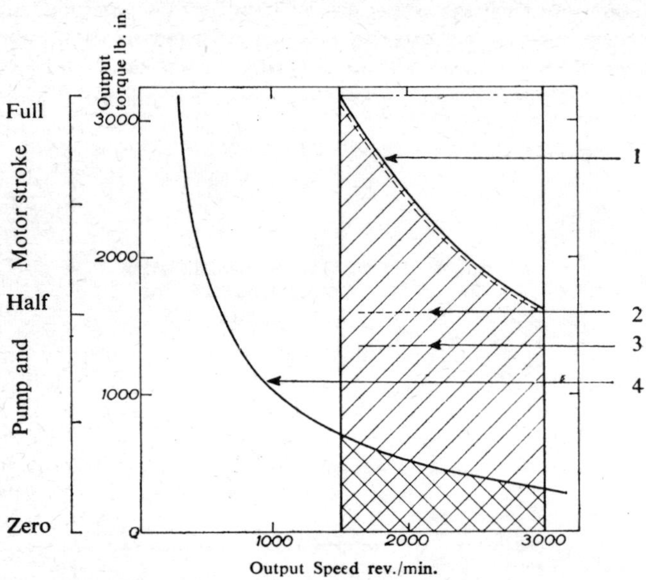

(B) Fixed delivery pump. Variable displacement rotor.

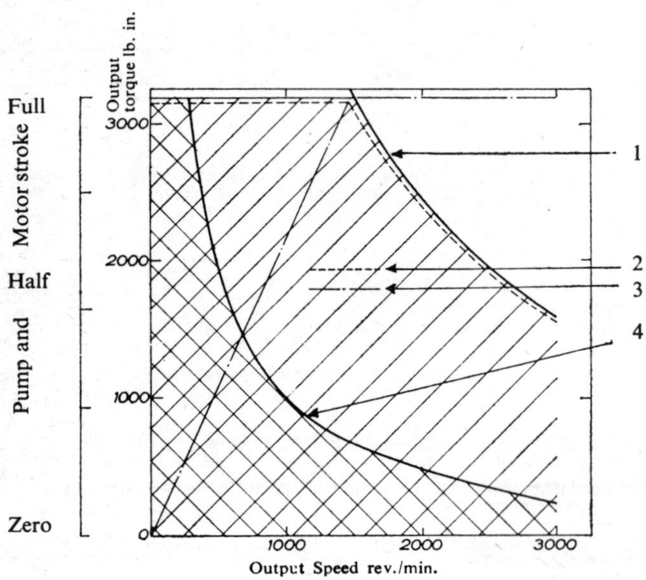

(C) Variable delivery pump. Variable displacement rotor.

Fluid Requirements

The basic requirements of a hydraulic fluid for hydrostatic transmissions are similar to those of any other high duty hydrostatic system, although in practice individual manufacturer's recommendations may vary from the use of straight mineral oils to special 'improved' oils with high viscosity indices, high shear stability and anti-wear additives. The majority normally recommend standard hydraulic oils, which already contain suitable additives, in the viscosity range of from 20 to 120 centistokes at 100 deg F, but specifically related to the pump and motor design. Basically the logical choice of viscosity is that affording the best possible compromise between efficient transmission, adequate lubrication and sealing under the actual service conditions involved.

The following are also primary requirements for hydrostatic transmission fluids and so will be further mentioned, although they duplicate the requirements in most respects, of any good quality hydraulic oil (eg see also Chapter 8).

Oxidation Stability

Resistance to oxidation determines the life of the oil and so oxidation inhibitors are normally specified for hydrostatic transmission fluids. This will then ensure maximum fluid life, the actual life depending largely on the operating temperature of the fluid in the system.

Anti-corrosion Properties

Anti-oxidation inhibitors ensure that the fluid will remain non-acidic for long periods. However this does not guard against the possibility of corrosion being initiated by water present in the fluid. Such rusting of ferrous metal components is particularly likely to occur during shut down periods and could seriously damage the transmission. The development of rust particles can also accelerate degradation of the oxidation inhibitor and promote oil oxidation when the inhibitor is exhausted.

Two additional methods of protection are thus desirable, under the heading of anti-corrosion inhibitors. One is an additive to improve the 'wetting' qualities of the oil so that the oil will maintain a protective film over the interior surfaces of the system, particularly when standing idle. The other is a rust inhibitor, added to the fluid. Either, or both, may be employed in the oil make-up.

Another factor may also show up, which is largely peculiar to hydrostatic transmissions – namely the fact that certain oil additives can affect stress corrosion. If there is evidence of this, the oil formulation may need changing. This is not a positive cure, for the initial fault in the case of stress corrosion may lie in the design of the system components themselves.

Demulsibility

This can be a significant factor in hydrostatic transmissions, particularly if the fluid can get contaminated with water. The addition of detergents normally provides satisfactory resistance to emulsification with absorbed water content up to about 2 per cent, although larger proportions of water may form stable emulsions which can cause operating problems, unless rapid separation is possible in a suitable part of the system. (See also Chapter 8).

Anti-wear Additives

Anti-wear additives are normally incorporated in all good quality hydraulic oils. Good anti-wear properties are normally required in any hydrostatic transmission, since close fitting rubbing elements are involved in the pump and motor. Apart from minimising wear on these parts, the elimination of wear products from the system is also useful in minimising conditions favourable to oxidation and emulsification of the oil.

It should be noted that even with anti-wear additives, wear products may well be generated during the initial running or 'break in' period and require removal by filtering, or even a fluid change. After a break-in period, additional wear normally occurs only slowly with a fluid formulated with anti-wear additives.

Cavitation and Air Entrainment

Cavitation in the pump or motor can cause erratic operation and, if persistent, damage or failure in these components. As far as the fluid characteristics are concerned, it is important that the oil used should have no volatile contents, nor be operated at such high temperatures that they can 'flash' on the suction side of the pump.

The entrainment of air can also promote cavitation since the bubbles formed by dispersed air can become focal points for cavitation. Air entrainment can be minimised by the use of anti-foam additives, which also control the formation of foam on top of the oil in open partsoof the circuit.

Compressibility and Bulk Modulus

The fluid pressure in transmission systems is commonly of the order of 5 000 psi (350 kg/cm^2), at which level the compressibility of mineral oil fluids may be appreciable. In general, this will not normally lead to trouble, but may reduce the overall efficiency of the system somewhat, since some energy is absorbed in compressing the fluid. In critical cases, however, the compressibility of the fluid has been responsible for promoting cavitation, particularly when using light oils at relatively high working temperatures. An oil fluid with a relatively high bulk modulus is thus to be preferred, although not necessarily essential – eg an aromatic base oil in preference to a paraffin-base oil.

It should be noted that the addition of viscosity index improvers to an oil tends to increase the compressibility of the oil slightly (ie reduce the bulk modulus). Again this is not usually significant, although the actual reduction may be of the order of 10 per cent.

Viscosity

Viscosity has already been mentioned in general terms. A primary requirement of many hydrostatic transmissions is that they may be called upon to operate over a wide range of ambient temperatures, starting up from cold. It is thus desirable to have a fluid viscosity which is not too high at normal ambient as to be excessively thick for cold starting. Apart from starting difficulties, in an extreme case component damage can result from an excessive initial viscosity in the system. The pump volumetric efficiency, on the other hand, will usually rise with an increase in fluid viscosity – seeFig 5.2.

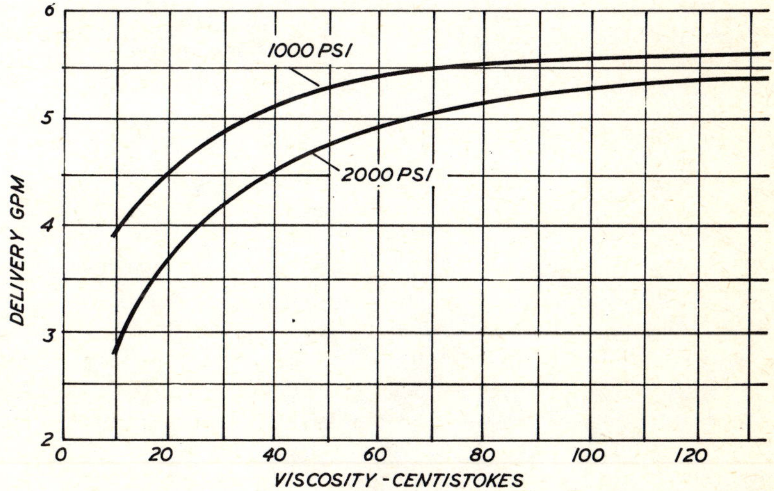

Fig. 5.2

Limits which have been specified are a maximum viscosity of 1 000 centistokes and a minimum viscosity (at maximum working temperature) of 10 centistokes. See Chapter 3 for the significance of these limits in the choice of a suitable oil viscosity.

A high viscosity index is obviously desirable in a hydrostatic transmission fluid which has to operate over a wide temperature range. This can be obtained by the addition of viscosity index improvers in the oil formulation. Such benefits are lost, however, if the shear stability of the resulting fluid is not good – ie the viscosity index improver breaks down and the oil reverts to its original lower viscosity index. Many hydrostatic transmission manufacturers place considerable significance on shear stability as a necessary fluid characteristic, particularly for high pressure systems working at relatively high fluid temperatures.

The effect of pressure on viscosity also cannot be ignored, especially in systems working at 5 000 psi or above (350 kg/cm^2 or above). Viscosity may be increased by 1.5 times at a pressure of 5 000 psi (350 kg/cm^2); and as much as 3 to 4 times at 10 000 psi (700 kg/cm^2). The increase with oils incorporating viscosity index improvers can even be greater than this at higher pressures in working systems because of the increased likelihood of partial shear breakdown – see also Table 5.1 and Chapter 1.

Compatibility

It goes without saying that the seals employed in the system must be compatible with the fluid chosen, or vice versa. There is, however, a subtle difference between seal compatibility and seal 'conditioning'. The latter refers to how soft and pliable the seal remains under continual immersion, and on standing. Seal/fluid combinations should be chosen on the basis of remaining well conditioned, even during periods of idleness.

Hydrokinetic Systems

Hydrokinetic machines transmit power by means of the velocity generated in a hydraulic fluid, with energy changes occurring as the fluid circulates round a totally enclosed circuit. Such devices may perform as couplings, where input and output torques are equal; or as torque converters. In the latter case a fixed reaction member is introduced between the driving and driven elements to change the direction of flow between the two elements. This results in variable torque-transmission characteristics. There is a *clutch point* or speed ratio at which input and output torques are equal. If the output is then slowed by an increased load, torque multiplication is achieved, rising to a maximum when the output is fully stalled. The two types – converter and coupling – can be combined in one design as concerter-couplings, having the advantages of both systems and maintaining a torque ratio of unity above the clutch point.

The elements of a hydraulic coupling are shown in Fig 5.3. The main components are an impeller fitted to the driving shaft and a runner fitted to the driven shaft, the impeller completely enclosing the runner to form a fully enclosed circuit for the hydraulic fluid (almost) filling the interior volume. In fact, the only significant flow is that around the torus sections formed by the impeller and facing runner. The remainder can be considered as 'make up' fluid, with the possibility of control via varying the amount of fluid, as is done in the case of the scoop-controlled coupling.

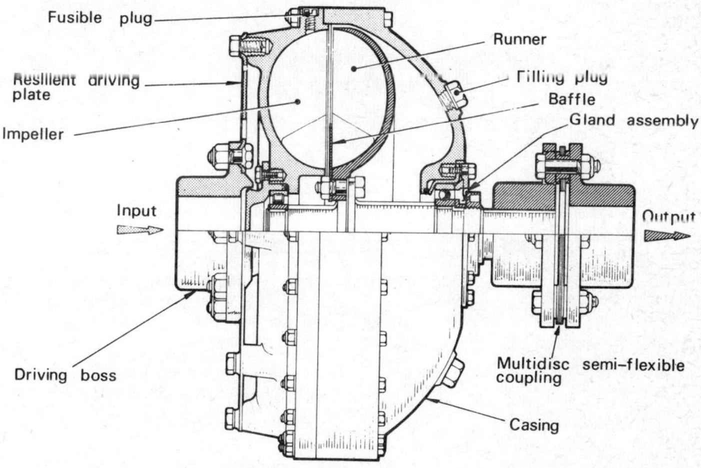

Fig. 5.3 Traction type fluid coupling with constant filling.

The torus path is asymmetric, the impeller outlet (radial width A) being greater than the runner outlet or impeller inlet (radial width B). Rotation of the impeller produces rotational flow in the torus with corresponding velocities vi and vo, where

$$V_i A = V_o B$$

For simplified analysis, the following equations then apply:

(1) Kinetic energy of flow across A $= \dfrac{m\, v_i^2}{2} = E_A$

(2) Kinetic energy of flow across B $= \dfrac{m\, v_o^2}{2} = E_B$

Since $A < B$, $v_i > v_o$, hence

(3) difference in kinetic energy $= E_A - E_B = E$

$$= \frac{\bar{Q}\,V_A^{\,2}}{2} - \frac{\bar{Q}\,V_B^{\,2}}{2}$$

where $\bar{Q}$ = rate of fluid flow, mass/second

V_A = linear velocity at A tangential to shaft rotation

V_B = linear velocity at B tangential to shaft rotation

(4) $HP = \frac{E}{k_p}$ where k_p is the horsepower constant for units employed

$$= \frac{\bar{Q}\,V_A^{\,2} - \bar{Q}\,V_B^{\,2}}{1100\,g}$$

where $\bar{Q}$ is in lb/second

V_A and V_B are in feet/second

$g = 32.2$ ft/sec²

(5) $\bar{Q} = k.f.V.a.\rho$

where k = constant for rationalisation of units employed

f = friction factor

V = fluid velocity along vortex paths

ρ = fluid density

a = cross sectional area of vortex path at impeller outlet

For V in ft/sec, a in square inches, and ρ in pounds per cubic inch; k = 12, yielding $\bar{Q}$ in pounds per second

The horsepower transmitted by the coupling can thus be determined from equations (4) and (5), provided the fluid velocity V can be established and a suitable friction factor determined. The latter can only be derived empirically and will vary with the flow conditions and slip. A typical value is 0.95 for laminar flow under low slip conditions, decreasing with increasing slip to approximately half that value at the onset of turbulent flow. The flow velocity along the vortex paths can be determined from first principles relative to the hydraulic head developed by the net centrifugal force generated by rotation.

Slip is an inevitable feature of a hydraulic coupling operating under practical load conditions, with a possible speed ratio of unity (no load on the driven shaft) to zero (driven shaft stalled). Input and output torque are identical, however. Maximum efficiency normally occurs with a slip of the order of 2.5 to 3 per cent, so the design of a coupling is normally proportioned to operate at this corresponding speed ratio under full load conditions.

For a given load the amount of slip is largely determined by the oil viscosity and thus slip will increase with fluid heating and consequent loss of viscosity. A higher viscosity oil will decrease initial slip but will generate more heat, will partially, or completely, nullify any gain in this respect. The main requirement is thus for a low viscosity oil and a high viscosity index, particularly for satisfactory starting characteristics at low temperatures.

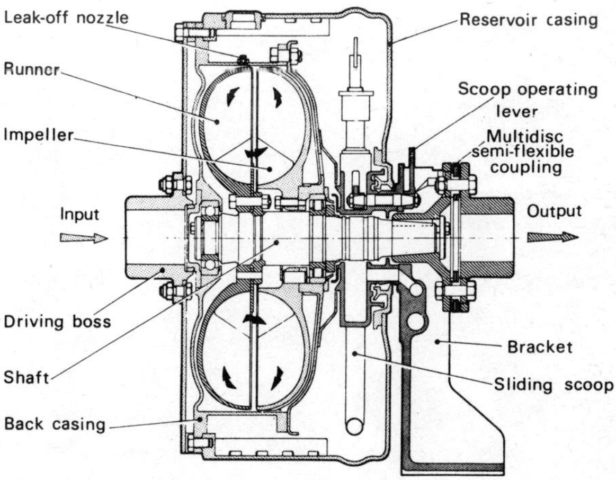

Fig. 5.4 Scoop control variable speed fluid coupling type SCR 4.

Fluid Requirements

The fact that a fluid coupling always runs with a certain amount of slip means that a proportion of the input energy is converted into heat, producing a rise in temperature of the fluid. In the stalled condition, all the input energy is dissipated in the form of heat, so that a very marked rise in fluid temperature may occur. In couplings of the constant filling type it may even be necessary to provide protection against overheating and excessive expansion of the oil by a fusible plug screwed into the casing.

Additionally, in the case of a simple coupling the design performance is specific to a given filled volume, and the effectiveness of coupling is largely governed by the quality of the fluid. At the same time high flow velocities are realised, with accelerations and decellerations, so that a low fluid viscosity is necessary in order to minimise fluid inertia and viscous losses (which would add to heating).

The requirements of a fluid for hydrokinetic devices therefore differs appreciably from that of hydrostatic systems. The choice is invariable (or almost invariable) a mineral oil with a viscosity of 10 to 30 centistokes at the operating temperature of the system, and of suitable high quality. Whilst it is desirable to have the lowest fluid viscosity, the lowest viscosity limit is usually set by lubrication requirements. Thus in the case of a coupling operating with relatively high slip or under other conditions promoting a relatively high working temperature for the fluid it may be necessary to employ a higher viscosity oil in order that a minimum viscosity of about 10 centistokes is realised at the working temperature.

In the case of automobile automatic transmissions a special additive is sometimes used in the fluid to eliminate noise which may be produced by the epicyclic gear brakebands and internal clutches as they engage and disengage at each speed change. This is generally known as an 'anti-squawk' additive. A similar additive is also used in limited-slip differentials for the same reason, where the unit is not incorporated in a common unit with the automatic transmission.

Scoop control couplings, in particular, generate a considerable amount of heat when driving at reduced speeds and thus require a higher viscosity oil to compensate for viscosity lossththan a typical traction coupling where operating temperatures are low and lubrication requirements are limited to a number of anti-friction bearings. Many torque converters, on the other hand, run for long periods at high torque ratios and require a low viscosity oil in order to avoid excessive generation of heat. Each individual kinematic machine design, in fact, represents a particular problem in oil viscosity requirements, as well as in oil quality.

In general transmissions of this type, and fluid couplings and straightforward torque converters, require a fluid with a high viscosity index to operate successfully over a wide range of temperatures; a low pour point to ensure satisfactory working at low ambient temperatures; and good anti-foaming characteristics.

TABLE 5.1
TYPICAL TEST DATA* – EFFECT OF PRESSURE ON VISCOSITY OF TRANSMISSION FLUIDS

FLUID	VISCOSITY (CENTIPOISE)			
	At atmospheric pressure		At 10 000 psi	
	100°F	200°F	100°F	200°F
Hydraulic transmission fluid (No. VI additive)	38.2	5.93	166	17.1
Hydraulic oil with VI additive	40.3	7.22	164	20.3
Hydraulic transmission fluid with VI additive	33.1	6.52	129	17.6

* Falling ball viscometer (American Oil Company)

TABLE 5.2
VISCOSITY RECOMMENDATIONS FOR PLAIN BEARINGS

Size: Diameter: ins.	Bearing Speed: rpm	Viscosity Requirements at Operating Temp.		
		32°F–90°F	90°F–140°F	140°F–200°F
Up to 1	Up to 3 000	SAE 10w	SAE 20	SAE 30
	Above 3 000	SAE 5w	SAE 20w	SAE 20
1 – 2	Up to 1 500	SAE 10w	SAE 20	SAE 30
	Above 3 000	SAE 5w	SAE 20w	SAE 20
2 – 3	Up to 1 000	SAE 10w	SAE 20	SAE 30
Above 3	Above 1 000	SAE 5w	SAE 20w	SAE 20
	Up to 750	SAE 10w	SAE 20	SAE 30
	Above 750	SAE 5w	SAE 20w	SAE 20

6 WATER HYDRAULICS

Water has the advantage of low cost and ready availability as a hydraulic fluid, but its numerous disadvantages normally preclude its use in modern hydraulic systems. These disadvantages are:-

(i) Lack of lubricating properties.

(ii) Promotion of rusting in contact with ferrous metal surfaces.

(iii) Limited temperature range for operating – freezing at only moderately low temperatures and generating high vapour pressure (and eventually boiling) at only moderately high temperatures.

(iv) Very low viscosity, (of circa 1 centistoke), which aggravates sealing problems, water-hydraulic systems, in general, being very prone to leakage.

Water is still employed in modem systems, however, in particular cases, namely –

(i) Very large systems of high capacity, particularly those served by a central accumulator and/or a central pumping station.

(ii) Particular systems where the fluid is discharged to waste, or a high capacity system has a high leakage rate.

(iii) As an inexpensive alternative to a fire-resistant fluid in hazardous situations, eg particularly in mining equipment.

(iv) For hydraulic burst tests where the fluid may be lost and the higher bulk modulus (lower compressibility) reduces the 'explosive' hazard.

(v) For systems that must be non-contaminating in the event of a burst or leakage. This is valid only when water is non-contaminating as far as the ambient is concerned, because of the inherently higher leakage rates with water as the pressurised fluid.

Lack of lubricating properties of water can, to some small extent, be improved by the addition of soluble oils. The main use of soluble oil additives, however, is to minimise corrosive attack, a 2 per cent ratio of oil to water generally giving maximum protection, although anything from 1 to 5 per cent oil additive may be used in practice. Non-petroleum base oil additives give somewhat better protection than petroleum-base additives for the same concentration. The chief objection to soluble oil additives is that they are prone to separation, either in the presence of dirt or contaminants in the system (eg corrosion products), or due to excessive concentration of additive.

Soluble oil emulsions are mostly used in large industrial hydraulic systems (eg with capacities of the order of 10–20 000 gallons). A typical application is in large forging presses.

Corrosion protection can be enhanced by applying protective coatings direct to ferrous metal surfaces or copper alloys, this generally being cheaper than the use of resistant metals. Typical treatment consists of heavy plated coatings, namely 0.005 inch (0.125 mm) chromium on steel; and 0.01 to 0.02 inch (0.25 – 0.5 mm) nickel on bronze – eg the 'Fescol' process.

The limited working temperature range of water can be enhanced by the addition of a freezing point depressant, typically glycerine or glycol. Water-glycerine mixtures were developed initially for marine hydraulic fluids, as an alternative to oils, with glycerine proportions ranging up to 50 per cent, depending on the minimum operating temperature required – see Table 6.1. Such additives are expensive, especially used in high proportions, and water-glycerine mixtures are more expensive than mineral oil fluids. They do have the additional advantages of being fire-resistant, possess a high viscosity index and a comparatively low coefficient of expansion, however. Water-glycol mixtures are used for the basis for a group of modern fire-resistant fluids, which are described in more detail in Chapter 8. Water-in-oil emulsions are similarly dealt with in that chapter, as distinct from soluble oil additives used with water in low concentrations and producing oil-in-water emulsions. The former are water-base or water-containing rather than true water fluids.

The kinematic viscosity of water is approximately 1 centistoke at 20 degrees C (0.0000108 ft^2/sec at 68 degrees F); and the range in actual viscosity is not very great from freezing to boiling point – see Table 6.2. The viscosity characteristics are not appreciably altered by soluble oil additives up to 5 per cent and so a true water fluid has very low viscosity at all usable temperatures – so much so as to aggravate sealing problems, as already mentioned.

The average compressibility of water is 0.336 per cent per 1 000 psi at 60 degrees F (10 deg C); or 49.5×10^{-6} per atmosphere at 60 degrees F (10 degrees C). This is consistent with a bulk modulus of 300 000 to 320 000 lb/ft^2.

Compressibility data are summarised in Table 6.3 over a pressure range up to 14 000 psi (1 000 kg/cm^2). In practice, the compressibility will also vary slightly with soluble contents and temperature. Water shows minimum compressibility at a temperature of around 122 degrees F (50 degrees C).

The vapour pressure of water can be a significant factor at higher temperatures, particularly when the water is pressurised by pumping. Vapour pressure data are summarised in Table 6.4.

TABLE 6.1
FREEZING POINT OF WATER-GLYCOL SOLUTIONS

% Glycol by volume	Ethylene Glycol °F	Ethylene Glycol °C	Propylene Glycol °F	Propylene Glycol °C
5	28	−2.2	29	−1.7
10	24.5	−5	26	−3.3
15	20	−7	23.5	−5
20	15	−9	19.5	−7
25	8	−13	14	−10
30	2	−17	8	−13
35	−6	−21	2	−17
40	−15	−26	−5	−20
45	−25	−32	−15	−26
50	−39	−39	−25	−32
55	−50	−50	−40	−45

TABLE 6.2
VISCOSITY OF WATER

Temperature °C	Temperature °F	Absolute Viscosity Centi-poise	Kinematic Viscosity Centistokes	Kinematic Viscosity Ft^2/SEC	Kinematic Viscosity in^2/SEC
0	32	1.793	1.79	.00001931	.00278
5	41	1.522	1.52	—	—
10	50	1.309	1.3	.00001412	.00203
15	59	1.142	1.14	—	—
20	68	1.008	1.008	.00001089	.00157
25	77	.893	.89	—	—
30	86	.800	.80	.000008681	.00125
40	104	.653	.65	.000007105	.001025
50	122	.549	.55	.000005995	.000864
60	140	.469	.47	.000005142	.00074
70	158	.407	.41	.000004484	.000645
80	176	.357	.36	.000003968	.000572
90	184	.316	.32	.000003534	.000508
100	212	.284	.28	.000003185	.000458

NOTE: the accepted standard viscosity for water (since 1953) is 1.0038 centistokes at 68°F.

TABLE 6.3A
COMPRESSIBILITY OF WATER AT 60°F

PRESSURE PSI	PRESSURE kg/cm^2	% REDUCTION IN VOLUME
1 000	70	0.33
2 000	140	0.66
3 000	210	1.00
4 000	280	1.33
5 000	350	1.66
6 000	420	2.00
7 000	490	2.33
8 000	560	2.66
9 000	630	3.00
10 000	700	3.30

NOTE: the compressibility of water is a minimum at about 50°C (122°F)

TABLE 6.3B
COMPRESSIBILITY OF WATER AT 20°C
(TABLE FIGURES SHOW DENSITY RELATIVE TO DENSITY AT ATMOSPHERIC PRESSURE)

Pressure psi	Relative Density
1 000	1.0033
2 000	1.0067
3 000	1.0094
4000	1.0125
5 000	1.0154
6 000	1.0183
7 000	1.0213
8 000	1.0238
9 000	1.0268
10 000	1.0290
11 000	1.0320
12 000	1.0346
13 000	1.0397

Pressure kg/cm^2	Relative Density
100	1.0046
200	1.0089
300	1.0132
400	1.0173
500	1.0213
600	1.0291
700	1.0290
800	1.0329
900	1.0365
1 000	1.0400

TABLE 6.4
VAPOUR PRESSURE OF WATER
(In mm of mercury)

Temp. °C	0	1	2	3	4	5	6	7	8	9
0	4.579	4.926	5.294	5.685	6.101	6.543	7.013	7.513	8.045	8.609
10	9.209	9.844	10.518	11.231	11.987	12.788	13.634	14.530	15.477	16.477
20	17.535	18.650	19.827	21.068	22.377	23.756	25.209	26.739	28.349	30.043
30	31.824	33.695	35.663	37.729	39.899	42.175	44.563	47.067	49.692	52.442
40	55.324	58.34	61.50	64.80	68.26	71.88	75.65	79.60	83.71	88.02
50	92.51	97.20	102.09	107.20	112.51	118.04	123.80	129.82	136.08	142.60
60	149.38	156.43	163.77	171.38	179.31	187.54	196.09	204.96	214.17	223.73
70	233.7	243.9	254.6	265.7	277.2	289.1	301.4	314.1	327.3	341.0
80	355.1	369.7	384.9	400.6	416.8	433.6	450.9	468.7	487.1	506.1
90	525.8	546.1	567.0	588.6	610.9	633.9	657.6	682.1	707.3	733.2
100	760.0	787.6	—	—	—	—	—	—	—	—

TABLE 6.5
SPECIFIC GRAVITY & SPECIFIC WEIGHT OF WATER

Temperature °F	Temperature °C	Specific Gravity	Specific Weight Lb/cu.ft*	Specific Weight lb/cu.ft ø
32	0	0.9999	62.42	62.35
39.2	4	1.0000	62.427	62.356
50	10	0.9997	62.41	62.34
60	16	0.9990	62.37	62.299 ✓
70	21	0.9980	62.30	62.23
80	27	0.9966	62.22	62.15
100	38	0.9931	62.00	61.93
200	93	0.9630	60.12	60.05

* in vacuo ø in air ✓ 0.03605 pounds per cubic inch

TABLE 6.6
THERMAL EXPANSION OF WATER

Temperature Range	Volumetric Expansion Coefficient per °C
5 – 10°C	5.3×10^{-5}
10 – 20°C	15.0 "
20 – 40°C	30.2 "
40 – 60°C	45.8 "
60 – 80°C	58.7 "

TABLE 6.7
SPECIFIC HEAT OF WATER

Temperature $^{\circ}C$	Calories/Gram/$^{\circ}C$
0	1.0094
5	1.0054
10	1.0027
15	1.0011
20	1.0000
25	.9992
30	.9987
35	.9983
40	.9982
45	.9983
50	.9987
55	.9992
60	1.0000
65	1.0008
70	1.0016
75	1.0024
80	1.0033
85	1.0043
90	1.0053
95	1.0063
100	1.0074

NOTE: the specific heat of water has a minimum value at about 37.5°C

7 OIL HYDRAULICS

The continued development and increasing application of industrial hydraulics has centred around closed circuit, or recirculatory low volume hydrostatic systems, where compactness of components can be realised by the use of high fluid pressures. Up until World War I, 1 000 psi (70 kg/cm^2) or even 500 psi (35 kg/cm^2) was regarded as a high pressure for small systems, but developments in aircraft hydraulics during and subsequent to the war years established the practicability and advantages of using much higher pressures. Thus aircraft hydraulics are now commonly 3 000 psi (210 kg/cm^2) systems, or even 5 000 psi (350 kg/cm^2) systems. For industrial hydraulics, 1 000 psi is now regarded as a 'medium pressure' system, with 1 500 psi (105 kg/cm^2) or 2 000 psi (140 kg/cm^2) the more usual choice.

Apart from the advantages gained with higher pressures – notably a reduction in the volume of fluid required and the size and weight of the components – the most significant features with higher pressures are that loads on unbalanced pump and machine components become greater, and closer fits are required to contain the fluid. This places a premium on the lubricating qualities of the fluid employed. For pumps to run at higher speeds or higher pressures without undue wear and loss of performance, they must receive adequate lubrication from the fluid they are handling (see also Table 7.1). For this reason the development of modern industrial hydraulics is virtually synonomous with the use of oil fluids which are, essentially, lubricants. At the same time, besides acting as a lubricant, the fluid must possess a number of other essential characteristics, such as good chemical stability and resistance to foaming, as well as protecting metal surfaces from corrosion – all these over a wide range of possible operating conditions. And the higher the system pressure, in general, the more exacting these requirements become.

TABLE 7.1
TYPICAL RECOMMENDED OIL VISCOSITIES FOR PUMPS
(For average working conditions at temperatures not exceeding 110°F, unless specified otherwise)

Maker	Name or Type of Pump	Oil Viscosity at 70°F*	
		Redwood No. 1 secs	Centistokes
Dowty	"Live-line"	300	74
Fielding & Platt	Radial	300	74
Fraser	"Monoradial"	300	74
	Rotary	300	74
Keelavite	Rotary	300	74
Oilgear	Type "F"—temp 30–45°F	300	74
	" 45–55°F	750	185
	" above 55°F	1 300	320
Stein Atkinson Vickers	Vickers—Detroit up to 1 000 psi	300	74
	up to 2 000 psi	750	185
Sundstrand	Rotary	300	74
Tangyes	Standard	300	74
Towler	"In-line" plunger pumps At normal temperatures	90–100	21–24
	At high temperatures	300	74
Vickers	Vane type—average service	300	74
	without control valve	500	124
	Below 500 psi with control valve	200	49
	without control valve	300	74
	Piston-type—average conditions	500	124
	below 500 psi	750	185
	Variable speed—temp 60°F	300	74
	Gear pumps 100°F	500	124
	120°F	750	185
	140°F	1 250	310

*Assuming operation under normal conditions

A straight mineral oil alone, regardless of its quality, cannot normally fulfill all the qualities required from a modern hydraulic fluid. For that reason the natural properties are nomally enhanced by up to five or six different types of additives – viscosity index improvers, anti-oxidants, anti-wear additives, anti-corrosion additives and foam inhibitors – see also Table 7.2. The necessity for these, and the proportions used, depend to some extent on the characteristics of the basic oil.

TABLE 7.2
OIL ADDITIVES

Additive	Hydraulic Oils	Crankcase Oils	Remarks
Oxidation Inhibitors	√	√	
Viscosity Index Improvers	√	√	
Metal Deactivators or Passivators		P	
Rust Inhibitors	√	P	
Anti-wear Agents	√	√	
Extreme pressure additives			Gear oils (EP lubricants) & Hypoid gear oils
Anti-foam Agents	√	P	Gearbox Oils
Emulsifying Agents			Soluble oils for water systems
Demulsifying Agents	√	P	
Detergent-Dispersants	P	√	
Pour Point Depressants	P	P	
Tackiness Agent			Processing machinery oils to avoid leakage
Colour Stabilisers or dyes	P	P	

√ = Usual additive P = Possible additive

Crude oil consists essentially of a mixture of paraffinic, naphthenic and aromatic hydrocarbons, the proportions being generally, or specifically, characteristic of the source of the crude. The first process in refining the crude is fractional distillation to extract and separate the various liquid fractions. These first fractions yield unrefined petrol, kerosene, diesel fuel, etc. The heavy residues are then further distilled, under vacuum, to yield a range of unrefined lubricating oils. These oils are then refined in various ways to produce refined basic oils from which motor oils, gear oils, hydraulic oils, etc, can be blended. The amount, and method, of refining depends largely on the nature of the crude.

Lubricating oils derived from paraffinic crude contain paraffin wax, which gives the oil a high pour point because of the readiness of the wax to solidify at lower temperatures. To improve the pour point the oil must be dewaxed, normally by dilution with a suitable solvent, chilling, removal of the wax by filtering and recovery of the solvent. Thus a 'dewaxed' oil refers, generally, to one of paraffinic origin which has undergone this treatment and thus should have a low pour point. The pour point of an oil may be further lowered if necessary by the addition of a pour point depressant (see later).

Further refining is carried out to remove unstable constituents of similar viscosity and volatility from the basic oil, which have distilled over with it. Solvent refining is generally preferred, using sulphur dioxide, furfuraldehyde, propane and similar solvents, which have a selective action in removing the undesirable components. An alternative is acid refining, which involves treatment of the oil with sulphuric acid, which attacks and destroys the less stable constituents, but less selectively. Remaining traces of free acid are then neutralised with an alkali, followed by treatment with an activated earth to clarify the oil.

De-asphalting is an additional form of solvent refining, used to remove asphaltic bitumens and similar heavy constituents, which could otherwise lead to deposit formation. In this case the solvent used is commonly propane. De-asphalting may also be done by distillation.

The end product is a 'straight' mineral oil, normally of paraffinic or naphthenic base (aromatic crudes being more desirable for petrols), which can be produced in a somewhat limited range of viscosities. These can be blended, as necessary, to produce a range of intermediate viscosities.

Individual characteristics and additives employed will now be considered in more detail under individual headings.

Viscosity

Any viscosity range quoted for hydraulic oils is largely nominal, particularly as it is the viscosity of the fluid at the working temperature which is of primary importance, whereas specification viscosities are quoted for normal ambient or standard (measurement) temperatures. The viscosity required is dictated largely by the requirements of the pump. Piston pumps, in general, tend to require a slightly higher viscosity oil than vane pumps; and gear pumps a somewhat higher viscosity than piston pumps. This is primarily due to the difference in heating experienced with different types of pumps and the difference in sealing lengths of the working components. As an overall rule, the choice of fluid viscosity would be based on the pump manufacturer's recommendations.

Viscosity-temperature characteristics vary with the type of basic oil, and method of refining. Paraffinic oils which have been solvent refined, normally have high viscosity indices; those refined in other ways a somewhat lower viscosity index. Naphthenic base oils have a lower viscosity index than paraffinic base oils. Thus typical values for the viscosity index of straight blended mineral oils may range from about 80–90 up to about 100–105. Any value above 80 is generally reckoned to be a high viscosity index. A straight naphthenic oil would have a viscosity index in the 40 to 80 range.

The viscosity index of any oil can be improved by the addition of long chain acryloid polymers (viscosity index improvers) in proportions of up to about 10 per cent. By this means the viscosity index can be increased up to about 160. Rather lower improvement is more normal, since a large improvement in VI can render the oil extremely shear sensitive, and high proportions of viscosity index improvers can have an adverse effect on other characteristics. Viscosity-temperature characteristics of some typical hydraulic fluids, with and without VI improvers, are shown in Fig 7.1.

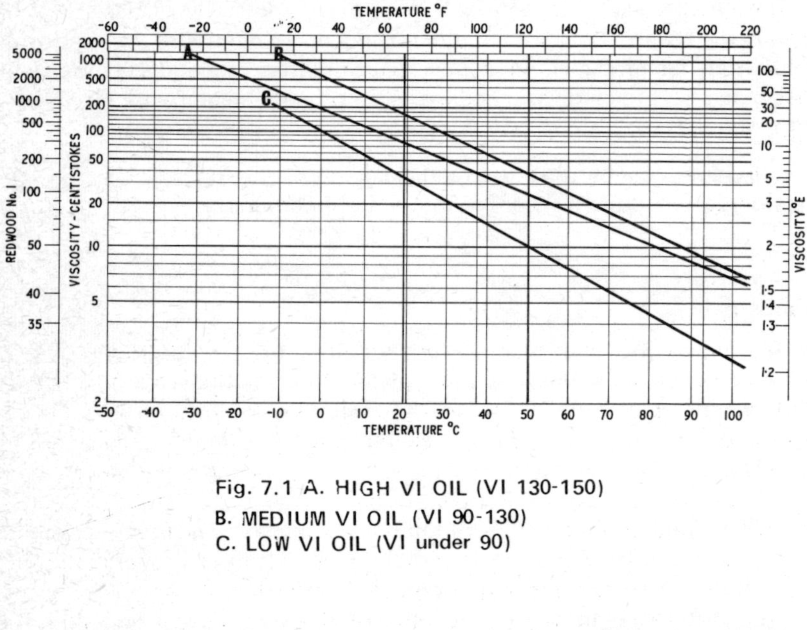

Fig. 7.1 A. HIGH VI OIL (VI 130-150)
B. MEDIUM VI OIL (VI 90-130)
C. LOW VI OIL (VI under 90)

Shear Stability

In a practical system, oil may be pumped through restricted areas at high pressure, generating high local velocities in the fluid and high shear stresses. The effect on viscosity index improvers is that the chain molecules can align themselves in the direction of flow, with temporary loss of viscosity; or under more severe shear, break up with permanent loss of viscosity. This is illustrated diagrammatically in Figs 7.2 and 7.3. The shear stability of a fluid can thus be assessed on the loss of viscosity experienced under specific flow conditions producing high shearing stresses in the fluid — see Chapter 11.

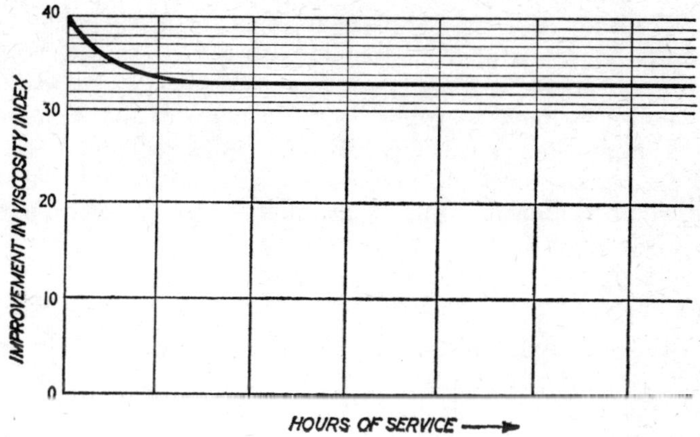

Fig. 7.2 Shear stability characteristic curve, showing initial loss followed by appreciably constant characteristics.

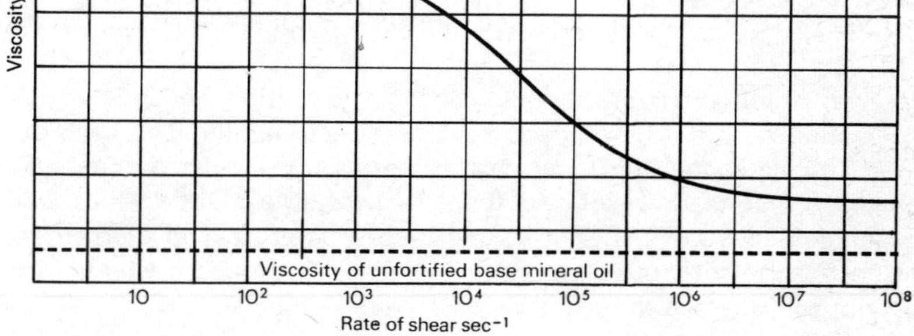

Fig. 7.3 Effect of rate of shear on very high V.I. oils.

High rates of shear are achieved at the tips of the gears in gear pumps, at the tips of the vanes in vane pumps, flow through restrictions and in rolling bearings. The shear stability of different VI improvers varies enormously. In all cases with a high VI oil, however, some initial permanent loss can be anticipated after a period of use. Temporary loss may then be experienced in parts of the system subject to high shear rates, as in Fig 7.3, but this is recovered and does not affect the behaviour of the oil in other parts of the system. In all cases the viscosity cannot fall below that of the unfortified oil at the same temperature.

Pour Point

It is a general characteristic of a straight oil with a high viscosity index (eg a paraffinic base oil), that its pour point is also relatively high. This may not be significant in the case of industrial systems, which are not subject to such low ambient temperatures on standing, as to prevent the oil from flowing on start up, but can be an important factor for systems subject to very low ambient temperatures, or working at very low temperatures – eg aircraft hydraulic systems at height. In that case pour point depressants are added to ensure that the oil remains in a fluid state at the lowest likely service temperature.

Chemical Stability

Chemical stability, as far as an oil fluid is concerned, is largely related to its affinity for oxygen, the resistance to oxidation being determined both by the type of oil and, more particularly, the service conditions. The natural stability of an oil is related directly to its degree of saturation – ie long chain (paraffinic) hydrocarbons react less readily with oxygen than the aromatic and unsaturated hydrocarbons.

Factors which promote oil oxidation are agitation and heating in the presence of air, which reaction can be accelerated by the presence of contaminants which can act as catalysts. Oxidation products which may be formed are of two types – insoluble sludge which can cause sluggishness of the system and also separate out to clog lines, filters and small clearance spaces; and soluble products which tend to thicken and degrade the oil and again cause sluggishness.

Useful oil life may be specified directly in terms of the degree of oxidation, ie the extent of contamination by oxidation products. This is characterised by a marked rise in oxidation after a period of time – Fig 7.4. The addition of anti-oxidants or oxidation inhibitors substantially extends this useful life, until all the inhibitor is used up and oxidation proceeds normally.

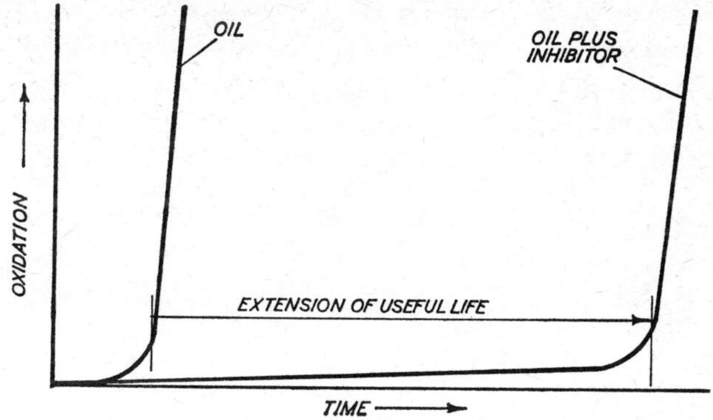

Fig. 7.4 Onset of oxidation with time, and extension of useful life provided by oxygen inhibitors.

Oxidation inhibitors work on the basis of a preferential attraction for oxygen, and thus prevent oil oxidation as long as they remain active. Protection cannot be extended indefinitely by increasing the proportion of inhibitor, however, for there is an optimum proportion of inhibitor beyond which no advantage is gained. This will vary with the type of oil, and inhibitor, but is normally of the order of 3 per cent.

Anti-corrosion Additives

Water and air are invariably present in oils (both being soluble to a limited extent in oil), whilst moisture may also be present from other sources, and is also formed by deterioration of oxidation inhibitors. Anti-rust or anti-corrosion additives are thus normally incorporated in a hydraulic oil formulation for additional protection of ferrous metal surface. These work on the basis of providing a strongly absorbed film on the metal surface, which prevents access of moisture and air to the surface. Alternatively, a wetting agent may be employed to ensure that the oil itself completely 'wets' the metal surfaces, although this is not a true anti-corrosion additive as such. Both types of additive may, in fact, be used in the same fluid, or a single additive perform both functions.

The selection of a suitable anti-corrosion additive must be done with care, since many of the most effective chemicals are reactive with, and can reduce the efficiency of, oxidation inhibitors and demulsifying agents. Some, in fact, are quite strong emulsifying agents.

It should also be noted that the anti-corrosion additives normally used in oils are anti-rust agents and thus may not be suitable for use with other metals. Anti-rust agents, for example, commonly attack zinc, as do many other additives. Zinc is also attacked by oxidation products in oils, with the formation of soaps. Zinc coated or galvanised surfaces should, therefore, not be used in oil hydraulic systems.

Anti-wear Additives

Although oils are generally excellent lubricants, full film lubrication may not be practical to obtain in certain components – eg at the tips of the vanes on vane pumps. This is because of the heavy pressure, or very small clearance, necessary to limit leakage. Boundary lubrication may only be marginal in such cases, leading to high wear rates. Thus for critical applications, very marked improvements in wear rate can be achieved with the addition of boundary lubricants or anti-wear additives to the oil.

The requirement is for an extreme pressure lubricant (eg phosphorus or sulphur type), but this poses difficult problems because of possible interaction with other additives. The selection of a suitable anti-wear additive must thus be undertaken with care, bearing in mind the other characteristics of the oil and the service conditions. Only a relatively few chemicals are suitable for this purpose, of which zinc alkyl dithiophosphate appears particularly favourable.

Water Separation

Moist air is invariably present, and drawn into the reservoir, with consequent condensation of water in contact with the oil. It is essential that the oil should remain separate from such water, so that it is not circulated through the system, ie the oil should have high demulsibility. If not, highly viscous water-in-oil emulsions could be formed in the reservoir and drawn through the system.

This desirable characteristic can be imparted by the addition of a demulsifying agent, although again this needs careful selection and balance. Extreme demulsibility, for example, can only be achieved at the expense of loss of anti-corrosion properties.

The presence of contaminants in the oil, such as finely divided iron oxides and hydroxides produced as wear products; and soaps produced by deterioration of the oil itself, will tend to produce persistent, stable emulsions with water. Anti-wear additives and antioxidants therefore, also play their part in improving the demulsibility of an oil.

Foaming

The solubility of air in hydraulic oils at normal temperatures and pressures is of the order of 8 to 10 per cent. More air can be dissolved at higher pressures, (approximately 2 per cent per 1 At pressure increase) and thus air would normally be released from solution at lower pressure points in the system – in practice, primarily in the reservoir. If the air liberated does not disperse quickly enough foaming can occur on the surface of the oil, which will prevent escape of further air which remains entrained in the fluid.

Most high quality hydraulic oils have inherent foam-resisting qualities, but these may be further improved by the addition of anti-foam additive. Such an additive works by modifying the size of the air bubbles produced during separation at lower pressures so that the air can escape readily without inducing foaming.

Entrained Air

To a large extent the ability of dissolved air to escape readily at low pressure points in the system, and of pressurised oil to retain more air in ~~solution,~~ reduces air entrainment problems, provided an excess of air is not already present in the system. This depends on the system being adequately vented, however. Residual air will then undergo a cyclic change, from being dissolved in solution at higher pressure points, and liberated at lower pressure points. Provided entrained air is absent on the pressurised side, no adverse performance effects are likely to be experienced. If, however, entrained air is fed to the pump, this will cause noisy and erratic operation, and this condition is obviously to be avoided.

This is primarily a system design rather than a fluid problem. As a general rule the possibility of air entrainment can be minimised by returning oil to the reservoir below the liquid level, so as to eliminate unnecessary churning; and locating the outlet in a suitable position so as to avoid vortex formation or erratic flow conditions which could entrap air and carry it into the delivery line. At the same time the reservoir design should allow maximum possibility for dissolved air to escape from the surface of the fluid.

Fluid viscosity has some significance in this respect. In a low viscosity fluid, the dispersing action of the pump often results in the formation of smaller air bubbles in the fluid reaching the reservoir, and thus taking longer to rise to the fluid surface and escape.

Toxicity

Mineral oils are not toxic and thus no special precautions are required in this respect. Castor-base fluids may, however, be toxic by nature on account of the solvents used.

Compatibility

Mineral oil fluids are compatible with a wide range of elastomers, although relatively few such materials are completely free from softening or swelling in contact with, or immersion in, oils. This subject is dealt with in detail in Chapter 9.

Castor-base fluids are more limited as regards compatibility. Originally they were produced for low temperature services to be used with natural rubber seals – natural rubber having a superior performance as regards hardening and embrittlement to most synthetic elastomers at low temperatures. It follows that the seals for these two types of oil fluids – mineral-base and castor-base – are incompatible, although some modern elastomers would be suitable for both services. Castor-base fluids are commonly employed for brake fluids and normally consist of a mixture of castor oil and methanol. The latter is a dilutant which does not affect the excellent lubricating properties of the castor oil. Castor oil and castor-base fluids are also particularly suitable for very high pressure hydraulics (see Chapter 1).

Low Temperature Stability

Low temperature stability may be significant in particular conditions, eg where a system has to stand idle at very low ambient temperatures, or work at very low temperatures, or for some reason hydraulic oil has to be stored at very low temperatures. In general, low temperature stability can be expressed in terms of absence of crystallisation, gelling, solidification or separation, etc, and a specified limit to turbidity, when exposed to a specified low temperature over a given period (usually a minimum of 72 hours).

Storage Life

All hydraulic oils should have a long storage life, although much will depend on the manner in which they are stored. Thus some segregation may occur if stored at low temperatures for an extended period, although the constituents may re-mix quite satisfactorily once the temperature of the oil is raised again. Should segregation be suspected then the oil should be brought to about 68 degrees F (20 degrees C) and allowed to re-mix for a period of 24 to 48 hours. Remixing can be aided by rolling the drums or stirring the contents.

Recommended storage temperature for most mineral oil fluids is not less than –4 degrees F (–20 degrees C) and not more than 86 degrees F (30 degrees C). Some segregation may occur if stored for extended periods at temperatures near to the lower limit.

During storage the possibility of water absorption should be minimised, such as by storing drums horizontally, and preferably under cover, so that no moisture can collect on dished ends. If the latter is allowed to happen there is always the possibility of water vapour being inhaled by expansion and contraction of the drum as filler cap seals are not necessarily impervious to water penetration under submerged conditions.

Care of Hydraulic Oils in Service

Where a good quality hydraulic oil of suitable viscosity is used in a system, a satisfactory life is normally ensured by the oil formulation, provided maximum working temperatures are not exceeded. Typical life figures to be expected are shown in Table 7.3.

TABLE 7.3
PROBABLE LIFE OF HYDRAULIC OILS (HOURS)

Service Conditions	Poor Quality Oil	Good Quality Oil	High Quality Oil
Very Good	2 000 – 3 000	10 000 – 16 000	Over 20 000
Good	1 500 – 2 500	8 000 – 10 000	12 000 – 16 000
Fair	1 000 – 2 000	4 000 – 5 000	8 000 – 12 000
Poor	500 – 1 000	2 000 – 4 000	4 000 – 8 000
Particularly Arduous	May be completely unsuited	500 – 2000	2000 – 4000

The greatest danger usually comes from solid contaminants, either already in the system in the form of abrasive dust particles, swarf, wear or corrosion products, etc; or introduced by incautious filling of the system. These contaminants need not necessarily detract from the oil life and performance, which is safeguarded by additives, but may materially affect the performance of the system by their clogging and blocking action. For this reason filters should be incorporated in the system, capable of removing and retaining solid particles down to the minimum size regarded as potentially harmful to the system.

The position of such filters should also be chosen with care, basic considerations being summarised in Table 7 4 Basically there is usually a choice of providing system protection by a high pressure filter on the pressure side filtering the main flow, or a low pressure filter on the return side at a point carrying the main flow, although there are obvious variations.

Wear products are most likely to be generated during the initial few hours running of a new system, and may account for quite a high proportion of ferrous metal contaminants being generated in this time, particularly in the case where the pump is of vane or gear type. Magnetic (plug) filters are particularly effective in trapping these initial wear products, and can be removed and cleaned after a run-in period. In critical installations a complete change of oil and filters is recommended after an initial run-in period.

TABLE 7.4 FILTER LOCATION

Position	Degree of Filtration	Remarks
Suction Line	25-40 microns	Must have low resistance to avoid pump cavitation. Low strength filaments suitable.
Main Pressure Line	down to 5 microns as required	May be by-passed for cold starts (not generally recommended). High strength elements desirable. High pressure case essential.
Individual Pressure feed lines	down to 3 microns as required	Gives maximum protection to components. High strength elements desirable. High pressure case essential.
Return Line	down to 5 microns as required	Similar requirements as for pressure filters but high strength case not required.
Relief Valve Return Line	down to 5 microns as required	As for return line filters.
Breathers	down to 20 microns	Fitted to reservoir or tank to eliminate airborne contaminants from system
External Filter	down to 10 or 5 microns if required	Employed during run-in period to assist in removal of 'built-in' contamination.
Magnetic Plug	ferrous particles only	Fitted to tank or reservoir (particularly useful during run-in period)
External Batch Filter	down to sub micron if required	Cleaning and reclaiming old oil in system. May also be employed to prefilter new oil.

Hydraulic Oil Specifications

The properties of individual hydraulic oils may be specified in detail (varying in degrees of completeness and coverage), although actual performance may vary slightly from batch to batch, although conforming to main requirements or limits. In the latter case performance figures quoted are generally referred to as 'typical inspections', derived from sample analyses; or 'typical physical characteristics', based on average performance data.

Various data for standard and various proprietary oil fluids are included here for specific reference.

TABLE 7.5

TYPICAL PROPERTIES OF MINERAL HYDRAULIC FLUID (DTD 585)

KINEMATIC VISCOSITY (centistokes)	4 500	450	45	10	4.5
at temperatures of	−60°C	−40°C	0°C	54°C	110°C
DYNAMIC VISCOSITY (centipoises)					
at gauge pressures (bars) of and at	0	200	400	600	1 000
temperatures of 30°C	14.4	20.1	28.2	39.0	71.7
60°C	7.35	10.0	13.6	17.8	29.5
90°C	4.62	6.16	7.91	10.0	15.5
Density at atmospheric pressure	.880	.854	.842	.830	.812
and temperature of	−40°C	0°C	20°C	40°C	70°C
at 20°C	.842	.851	.858	.864	.876
and gauge pressure (psi) of	0	2 000	2 000	6 000	10 000
COEFFICIENT OF THERMAL EXPANSION (per °C)	.00083	.00075	.00069	.00064	.00057
at gauge pressures (psi) of	0	2 000	4 000	6 000	10 000
COMPRESSIBILITY at 15°C (% vol)	1.1	2.4	4.3	7.0	12.7
and at pressures (psi) of	2 000	5 000	10 000	20 000	50 000
at 14 200 psi (% vol)	4.4	4.6	4.9	5.2	6.0
and at temperatures of	0°C	10°C	20°C	30°C	50°C
VAPOUR PRESSURE (mm/kg)	.0025	.05	2.5	50	165
at temperatures of	20°C	50°C	100°C	150°C	175°C
SPECIFIC HEAT (cal/gm/°C)	.42	.46	.48	.53	.59
at temperatures of	−20°C	25°C	50°C	100°C	175°C
SPONTANEOUS IGNITION TEMP(°C)	335	270	235	165	155
at gauge pressure (psi) of	0	50	200	500	1 000
FLASH POINT (open cup) 200°F	POUR POINT −90°F		VISCOSITY INDEX 200		
BULK MODULUS 184,000 psi	THERMAL CONDUCTIVITY at 15 deg.C		.000324 cal/cm sec deg.C.		

TABLE 7.6

TYPICAL CASTOR-BASE HYDRAULIC FLUID
(SAE 70–81 Heavy Duty Fluid)

Viscosity (Kinematic), cs	
at −40°F, max	1 800
at 130°F, min	4.0
Cold Test (A), 6 days at −40° F	Shall flow in 5 secs. No stratification or precipitate.
Cold Test (B), 6 hrs at −60°F	Shall flow in 5 secs.
Boiling Point, °F, min.	300
Boiling Point, after 2 hrs at 295°F, min	295
Flash Point (Cleveland) °F, min	145
Water Tolerance (100 ml fluid 3½ ml water)	
24 hrs at 140°F &	Shall flow in 5 secs.
24 hrs at −40°F	No stratification.
	Light precipitate allowed.
Corrosion Test, loss in mg per sq cm surface, max	
Tinned iron	.2
Steel	.2
Aluminium	.1
Cast Iron	.2
Brass	.5
Copper	.5
Neutrality	
Before corrosion test	pH 7–11
After corrosion test	pH 6–11
Rubber Swell, 1¼" cup, 120 hrs. at 158°F, inches	.005–.05
Compatibility (50/50 mixture with other fluids)	
24 hrs at 140°F	No stratification or precipitate.
24 hrs at −40°F	No stratification or precipitate.
Evaporation, 48 hrs at 210°F, % wt, max	80
Lubrication (Stroking test, 150 000 strokes at 158°F)	Condition of all components satisfactory.
Residue & Corrosion	No rust or Corrosion of metal parts.
	No hard, dry or gummy residue.

PROPRIETARY FLUIDS

BP ENERGOL HYDRAULIC OILS AND FLUIDS

Grade	Specific Gravity at 60°F	VISCOSITY Redwood No. 1 Seconds at 70°F	100°F	140°F	200°F	210°F	Centistokes at 100°F	210°F	Closed Flash Point °F	Pour Point °F
BP Energol HL 40	0.870	79	52.5	40	33	32.5	14.4	3.01	305	–50
BP Energol HL 50	0.873	163	84	51	37	36	22.6	4.20	360	–20
BP Energol HL 65	0.875	300	133	65	41	39	32.7	5.18	410	–20
BP Energol HL 80	0.878	440	183	80	44.5	42	44.9	6.28	420	–20
BP Energol HL 100	0.880	650	253	100	49.5	46	61.6	7.65	425	–20
BP Energol HL 150	0.883	1150	415	150	60	55	101.7	10.62	440	–20
BP Energol HL 175	0.885	1450	500	175	65	59	126.0	12.24	460	–10
BP Energol SHF100	0.875	420	205	98	53.5	50	50.5	9.1	415	–30
BP Energol SHF150	0.876	675	320	150	69	64	78.3	13.6	420	–20
BP Energol SHF11	1.078	275	140	74	–	–	34	–	–	–60
BP Energol SHF13	0.960	850	370	160	–	–	90	–	–	Below 0
BP Energol SHF0303	1.280	440	150	60	37	35.5	37	3.96	–	– 5
BP Energol SHF0305	1.330	1000	270	80	39.5	37.5	66	4.8	–	5

CASTROL HYSPIN HYDRAULIC OILS

Hyspin hydraulic oils are mineral-oil base with additives to inhibit oxidation and sludge and lacquer formation. Demulsifying agents and foam supressants are also included. The lightest oil, Hyspin 40, has a viscosity of 40 seconds Redwood No 1 at 140 degrees F and the heaviest, Hyspin 175, a viscosity of 175 seconds Redwood No1 at 140 degrees F. The specification number of the oil gives the normal viscosity (Redwood No 1) at 140 degrees F.

Hyspin AWS Grades

Property	Test method	Fluid			
		Hyspin AWS 10	Hyspin AWS 15	Hyspin AWS 22	Hyspin AWS 32
Specific gravity 60/60 °F	IP 160 – ASTM D1298	0.870	0 875	0.886	0.873
Kinematic viscosity – cS at 100 °F at 210 °F	IP 71 – ASTM D445	10.4 2.51	14.6 3.11	23.3 4.2	34.5 5.42
Viscosity index	IP 226 – ASTM D2270	75	70	86	100
Pour point – F	IP 15 – ASTM D97	–50	–40	–40	–20
Flash point (C.O.C.) – °F	IP 36 – ASTM D92	350	345	370	450

Aniline point – °C	IP 2	88.7	84.9	87.0	98.3
Seal compatibility index	A.H.E.M. Nitrile Rubber Swell Test	25	21	17	12
Foaming characteristics tendency/stability – mls/mls.					
at 75 °F	IP 146 – ASTM D892	50/NIL	50/NIL	50/NIL	50/NIL
at 200 °F		50/NIL	50/NIL	50/NIL	50/NIL
at 75 °F after 200 °F test		75/NIL	75/NIL	75/NIL	75/NIL
4 ball load test					
Incipient seizure load (I.S.L.) – Kgs.		55	95	95	105
Mean wear scar diameter at I.S.L. – m.m.		1.78	2.19	2.11	2.15
Weld point – kgs		130	160	140	160
4 ball load test					
Mean wear scar diameter – m.m.	1 hr at 30 kg.	0.47	0.33	0.31	0.30
Copper corrosion	IP 154 – ASTM D130 —168 hrs at 100 °C	1A	1A	1A	1A
Rust preventing characteristics –					
Distilled water	IP 135 – ASTM D665	Pass	Pass	Pass	Pass
Salt water		Pass	Pass	Pass	Pass

Hyspin AWS Grades

Property	Test method	Fluid			
		Hyspin AWS 46	Hyspin AWS 68	Hyspin AWS 100	Hyspin AWS 150
Specific gravity 60/60 °F	IP 160 – ASTM D1298	0.878	0.881	0.885	0.887
Kinematic viscosity – cS at 100 °F at 210 °F	IP 71 – ASTM D445	46.0 6.4	61.4 7.8	94.0 10.2	150 14.0
Viscosity index	IP 226 – ASTM D2270	100	100	100	97
Pour point – °F	IP 15 – ASTM D97	–5	–5	0	0
Flash point (C.O.C.) – °F	IP 36 – ASTM D92	460	470	460	460
Aniline point – °C	IP 2	102.0	103.1	110.5	114.9
Seal compatibility index	A.H.E.M. Nitrile Rubber Swell Test	11	8	6	5

Foaming characteristics tendency/stability – mls/mls –					
at 75 °F	IP 146 – ASTM D892	50/NIL	50/NIL	50/NIL	50/NIL
at 200 °F		50/NIL	50/NIL	50/NIL	50/NIL
at 75 °F after 200 °F test		75/NIL	75/NIL	75/NIL	75/NIL
4 ball load test					
Incipient seizure load (I.S.L.) – kgs.	—	100	110	100	105
Mean wear scar diameter at I.S.L. – m.m.		2.07	2.17	2.02	2.02
Weld point – kgs		200	210	200	240
4 ball wear test					
Mean wear scar diameter – m.m.	1 hour at 30 kg.	0.30	0.30	0.30	0.31
Copper corrosion	IP 154 – ASTM D130 168 hours at 100 °C	1A	1A	1A	1A
Rust preventing characteristics –					
Distilled water	IP 135 – ASTM D665	Pass	Pass	Pass	Pass
Salt Water		Pass	Pass	Pass	Pass

Hyspin AWH Grades

Property	Test Method	Fluid	
		Hyspin AWH 15	Hyspin AWH 46
Specific gravity 60/60 °F	IP 160 – ASTM D1298	0.876	0.876
Kinematic viscosity – cS			
at 100 °F	IP 71 – ASTM D445	15.9	44.2
at 210 °F		3.78	8.82
Viscosity index	IP 226 – ASTM D2270	142/143*	196/197*
Shear stability			
Viscosity drop after:			
1 cycle – % at 100 °F	Diesel Injector Rig	4	12
1 cycle – % at 210 °F		6	12
100 cycles – % at 100 °F		12	23
100 cycles – % at 210 °F		12	25
Pour point – °F	IP 15 – ASTM D97	– 60	– 25

Flash point (C.O.C.) – °F	IP 36 – ASTM D92	330	400
Aniline point – °C	IP 2	81.5	95.5
Seal compatibility index	A.H.E.M. Nitrile Rubber swell test	23	13
Foaming characteristics tendency/stability – mls/mls at 75 °F at 200 °F at 75 °F after 200 °F test	IP 146 – ASTM D892	50/NIL 50/NIL 75/NIL	50/NIL 50/NIL 75/NIL
4 ball load test Incipient seizure load (I.S.L.) – kgs. Mean wear scar diameter at I.S.L. – m.m. Weld point – kgs	—	75 1.97 210	70 1.84 220
4 ball wear test Mean wear scar diameter – m.m.	1 hour at 30 kgs.	0.33	0.32
Copper corrosion	IP 154 – ASTM D130 168 hours at 100 °C	1A	1A
Rust preventing characteristics Distilled water Salt water	IP 135 – ASTM D665	Pass Pass	Pass Pass

DUCKHAMS HYDRAULIC OILS

Zircon

A range of high quality inhibited hydraulic oils designed to cater for the most arduous and precise operating conditions, where the hydraulic medium is required to function efficiently for long periods, without renewal, at elevated temperatures and pressures.

Contains Anti-scuff additives to the A.P.I. MS level for anti-wear properties.

Zircon 1

Extra light, for small high speed multi-plunger type hydraulic pumps.

Zircon 2

Very light viscosity low pour point hydraulic oil for micro clearance systems and pumps.

Zircon 3

Light oil with a low pour point. Mainly for machine tool hydraulic systems.

Zircon 4

Medium oil for most types of machine tool applications where working temperatures do not exceed 100 deg F.

Zircon 5

Medium heavy oil. As above, but where temperatures are likely to reach 130 deg F.

Zircon 6

Heavy oil for temperatures exceeding 130 deg F. Also for fluid flywheels, torque converters, etc.

Zircon 7

Heavy oil for units applying high pressures at slow speeds. Also for vane type pumps.

Zircon 8

Extra heavy oil. As for Zircon 6 but when temperatures are above 150 deg F.

Zircon Q

A range of two special hydraulic oils with very high viscosity indices, inhibited against oxidation and foaming, specially developed to overcome time cycle variations in precise machine tools.

Zircon Q4

Medium viscosity oil for temperatures up to 130 deg F.

Viscosity index is in excess of 130.

Zircon Q6

Heavy oil for use where operating temperatures exceed 130 deg F.

Viscosity index is in excess of 130.

Note: Throughout the ZIRCON and ZIRCON Q ranges, the first suffix digit corresponds with the P.E.R.A. system of viscosity classification. The second digit 2 signifies the absence of mild EP additives.

NUTO (Esso) RANGE OF HYDRAULIC OIL FLUIDS

	Specific Gravity 60/60°F	Closed Flash Point °F	Open Flash Point °F	Pour Point °F	Viscosity – centistokes		Viscosity Index
					100°F	210°F	
NUTO H.32	0.873	190	215	–40	5.1	1.8	90
NUTO H.36	0.872	240	265	–35	14.1	3.1	90
NUTO H.40	0.878	280	310	–30	22.1	4.1	90
NUTO H.44	0.880	360	360	–25	33.1	5.3	95
NUTO H.54	0.888	400	430	–20	63.1	7.8	97
NUTO H.64	0.899	425	455	–25	105.0	10.7	92

LORCO HYDRAULIC OILS

LORCO HYDRAULIC OILS contain the Lorco four-fold treatment giving improved oxidation stability, anti-foam, anti-rust, anti-wear and anti-corrosion. They also meet Vicker's requirements for service M.S. lubricating properties.

Lorco M.4215 hydraulic oils have been tested and approved by Lucas Industrial Equipment for use in their hydraulic pumps.

Hydraulic Grade Viscosity Number		L 26	M 28	MH 31	H 33	XH 35	XX 37	Y 39	YY 41	Z 45
Method of measurement	**Property**									
IP71/62	210°F	5.4	6.6	8.4	9.7	11.7	14.9	18.0	23.3	32.7
Kinematic Viscosity (Centistokes)	100°F	33.0	45.5	68.0	85	112	160	218	340	560
IP70/62	70°F	290	440	700	900	1250	1850	2600	4250	8000
	100°F	133	186	270	340	450	650	850	1350	2250
Redwood Viscosity (Seconds)	140°F	67	82	107	130	162	225	280	400	625
	200°F	41.5	45.5	51.5	56	63	77	88	112	157

Viscosity Index			105	105	104	103	102	100	98	98	96
IP15/60(B) Pour Point †		°F	0°F	0°F	0°F	0°F	0°F	0°F	+15°F	+15°F	+15°F
Specific Gravity		60°F	0.870	0.871	0.877	0.883	0.888	0.891	0.895	0.899	0.903
IP84/58 / IP35/42 Flash Point	°F	Closed	405	420	450	460	495	505	520	530	550
		Open	430	460	485	490	515	525	535	550	575
IP2/81/A Aniline Point		°C	100	102	103	105	108	109	112	116	119
IP146/55T	Foamings:		(Foam volume ml)								
Tendency	@	75°F	10	10	10	10	10				
Stability	@	75°F	nil	nil	nil	nil	nil				
Inorganic Acidity		KOH/gm	nil	nil	nil	nil	nil	nil	nil	nil	nil

MOBIL D.T.E. OILS

The Mobil D.T.E. named and double lettered oils have a high viscosity index and are specially fortified for protection against rust. Other additives include demulsifying agents, anti-foaming agents and film strength improvers. Mobil D.T.E. 24, 25 and 26 are recommended specially for use in Vickers square vane type, VHO (high performance), intra-vane and other heavy duty hydraulic pumps. These pumps are used on both outdoor mobile equipment and also on machine tools. Mobil D.T.E. 24, 25 and 26 possess all the desirable characteristics of the Mobil D.T.E. named oils plus improved film strength properties to meet the severe operating conditions imposed by the moving parts in these heavy duty hydraulic pumps.

	Mobil D.T.E. Oils											
	Light	Medium	Heavy Medium	Heavy	Extra Heavy	BB	AA	HH	LC	24	25	26
Specific Gravity	0.870	0.875	0.880	0.890	0.895	0.900	0.900	0.905	0.870	0.875	0.880	0.880
Pour Point °F Max.	–20	20	20	20	25	25	25	25	–20	20	20	20
Flash Point C.O.C. °F Min.	395	400	410	420	420	440	460	520	350	395	395	400
Viscosity: c/s100°F	31	48	63	93	138	197	388	540	35	31.5	48	64
210°F	5.3	6.9	8.1	10.2	13.0	16.5	26	31	7.4	5.26	6.6	8.1
S.S.U.100°F	150	225	295	430	640	915	1800	2500	164	150	225	300
210°F	43	49	53	60	70	84	125	150	50	43.5	48	52
Redwood No. 1 70°F	280	460	650	1015	1625	2400	5100	7500	270	280	480	660
140°F	65	86	106	140	190	260	460	600	77	65	84	105
200°F	41	46	51	58	69	85	130	158	47	41.5	45	50
Viscosity Index Min.	103	100	97	95	95	95	95	95	190	95	95	95

SHELL TELLUS OILS

Viscosity

The new Shell Tellus Oils are available in a restricted range of viscosity grades which covers the known and readily foreseeable requirements of extra heavy duty hydraulic systems.

Demulsification Propensities

In both the demulsibility test ASTM 1401 and the steam emulsion number test IP19 the new Shell Tellus Oils show themselves to be somewhat superior to the grades they replace, that is to say they have greater power to shed entrained water.

Foaming Tendencies

The new Shell Tellus Oils have the same very high power to disperse foam as their predecessors.

Protective Properties

The outstanding degree of protection against corrosion associated with Shell Tellus Oils is maintained in the new grades.

Copper Strip Corrosion Test IP 154/59

Up to a temperature of 160 deg C, that is far above the temperatures ever reached by hydraulic media, the new Shell Tellus Oils are less reactive to copper and its alloys than conventional hydraulic oils. Above this temperature the oils become somewhat active to yellow metals. This is no disadvantage in general hydraulic engineering and arises from the presence of the anti-wear additives.

Oxidation Stability

Comparison of performance by the Turbine Oil Stability Test IP 157/62 shows a remarkable improvement in oxidation stability even over the very high standard associated with the former Shell Tellus Oils. It is a characteristic of the additives used in the improved grades that they show initially a false high reading in the standard test for acidity (IP 1/60). This falls very rapidly to a very low value which is maintained for an extremely long life before normal oxidation causes a final slow rise.

The Shell Tellus Range – Typical Data

Grade	S.G. at 60°F	Flash Point Closed °F	Pour Point °F	Viscosity Redwood No. 1 seconds at			
				70°F	100°F	140°F	200°F
Shell Tellus							
Oil 11	·836	145	—40	50	40	35	
,, 15	·865	310	—20	85	56	42	
,, 23	·873	340	—20	225	105	57	
,, 27	·875	410	—20	320	140	68	42
,, 29	·878	420	—20	510	210	89	47
,, 33	·880	420	—20	725	280	108	51
,, 41	·885	450	—10	1,250	460	160	62
,, 69	·888	460	—10	1,700	585	200	70
,, 72	·890	460	—5	2,500	875	280	90

8 SPECIAL FLUIDS

Special-purpose fluids are those developed for systems where the use of a mineral-oil fluid implies certain limitations, or hazards. Commercially the latter is the predominant factor responsible for the development of special fluids which are non-inflammable, or fire-resistant, under the general grouping of 'water-based' and 'synthetic', respectively. The distinction between a 'non-inflammable' and 'fire-resistant' fluid is not exact under such definition, since only water is a true non-inflammable fluid. Most water-based fluids are fire-resistant rather than non-inflammable, with a strong 'snuffing' action. Synthetic fluids range in characteristics from being substantially non-inflammable to having limited flammability. All such special fluids are thus most accurately classified as being fire-resistant.

There are also special applications in which synthetic fluids may show enhanced performance over oil fluids or water-base fluids — particularly where systems have to work at extremes of high, or low, temperature. Satisfactory low temperature performance, however, can usually be obtained with water-glycol solutions at much lower cost. Since synthetic fluids cost from 3 or 4 times as much as oil fluids, upwards, they are not considered as alternatives, but essentially solutions for special duties only. There are possible exceptions, eg for a system which has a low fluid volume and thus the greater cost of the fluid is not significant compared with specific advantages which may be offered; and for pure lubricants.

Fire-resistant fluids are an attractive alternative to oil fluids for high pressure industrial hydraulic systems operating in close proximity to naked flames or high temperature sources, or specifically in hazardous ambients where fire and/or explosion risk must be reduced to a minimum. The other important field of application is for aircraft hydraulic systems, where system pressures are normally higher than in industrial practice, and the consequences of fire even more drastic. The primary risk, in all cases, is that failure of, or damage to, a pressurised system can lead to the release of a fine spray of fluid which can be ejected a considerable distance; and in the case of an oil fluid this spray is highly combustible.

Whilst the advantages of using a fire-resistant fluid in such cases are obvious, from the safety point of view, the adoption of such fluids inevitably involves disadvantages, notably in the respect of higher cost, some loss of lubricating properties, and problems of seal compatibility, compared with mineral oil fluids. Their application is, therefore, still far from universal in systems even where a fire risk is present, although they may be mandatory in others. Also no 'standard choice' has been established for fire-resistant fluids in the same way that a mineral oil is a standard choice for a conventional system. Rather the various types of fire-resistant fluids continue to be used in individual spheres, and may be regarded as competitive in others.

The main types of fire-resistant fluids are:-

(i) Water-in-oil emulsions
(ii) Water-glycol mixtures
(iii) Phosphate ester fluid
(iv) Phosphate ester mixtures
(v) Chlorinated hydrocarbons
(vi) Silicones

General properties and characteristics of these various fluids are summarised in Table 8.1. A number of the more sophisticated synthetic fluids have been omitted from the tabulation since these are not, at present, commercial fluids and remain extremely costly; also their potentialities have not yet been fully developed. Silicones are included, as these are essentially fire-resistant fluids and were originally developed as such, although they again are extremely costly and their main application is as high temperature fluids (see later).

TABLE 8.1
COMPARATIVE PROPERTIES OF HYDRAULIC FLUIDS

Fluid	Specific Gravity (Typical)	Poor Point (Typical) °F	Max. Service Temp. °F	Acidity (Neutral-isation)	Water Content %	Comparative Lubricity (Bearing* load factor)	Vapour Pressure	Viscosity Index	Flash Point (Typical) °F	Auto Ignition Point °F	Com-parative Weight
Water	1.0	32	150	–	100	–	High	V. High	None	–	115
Mineral oils	0.865	0	200	0.06	0.02	1.0	Low	High	430	735	100
Water-in-oil Emulsions	0.92–0.94	–	150	0.05	40 min.	2.0–3.0	High	High	None	815	110
Water-glycol	1.06	down to −75	150	Alkaline	30–40	4.0	High	V. High	None	1 195	120
Phosphate Ester	1.275	0	300	0.02	0.03	1.2	Low	Low	500	1 200	125
Phosphate Ester Blends	1.15–1.30	10–15	300	0.03–0.05	0.03	1.2	Low	Low	475	1 200	130
Chlorinated Hydro carbons	1.43	30 below	300	1.10	0.02	1.2	Low	Low	420	1 200	160
Silicones	0.93–1.03	−70	600–700	–	–	–	Low	High	210–300	400–450	100

*See also Table 8.3

Water-in-oil Emulsions

These are basically water-containing oil fluids, the water content of which may range up to 60 per cent. The oil content provides reasonable lubricity and good corrosion protection; and a reasonably high viscosity (typically of the order of 50 to 60 centipoises at 70 degrees F (20 degrees C) with a water content of 40 per cent). Viscosity index is also high, and usually of the order of 140.

Such fluids are normally formulated from low viscosity mineral oils in which the water content is dispersed in the form of fine droplets of around 2μ diameter, together with additives to maintain a stable emulsion and inhibit oxidation (of the oil content) and foaming. The viscosity realised is a virtual rather than a true figure since the emulsion is a non-Newtonian fluid and is thus dependent on the rate of shear.—Fig. 8.1. Thus with high shear rates, such as could occur in close clearance spaces, the fluid may be subject to shear breakdown and permanent loss of viscosity. Subject to continuous high shear the emulsion could break down completely, with the viscosity reverting to that of the oil content. This places limitations on the application of water-in-oil fluids for use with certain components, eg with vane pumps, although the stability of modern emulsions has been

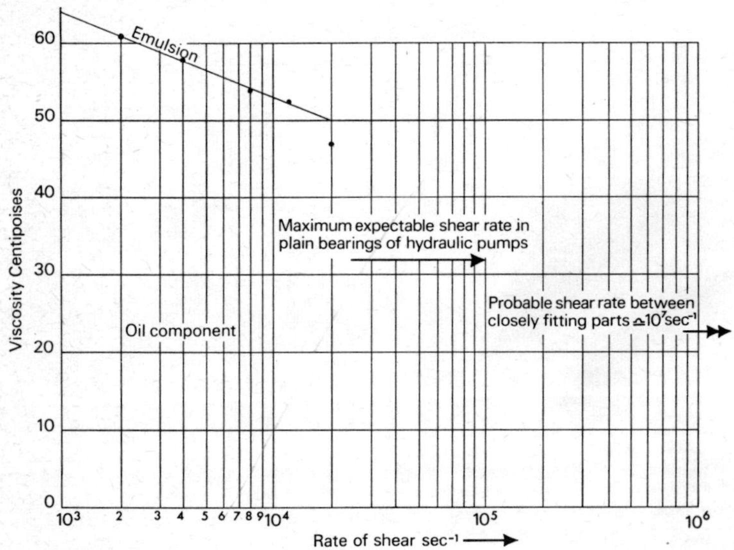

Fig. 8.1 Shear breakdown of typical water in oil emulsion at temperature of 100^0 F (38^{0}C)

considerably improved by the use of suitable additives. The latter also helps offset another inherent limitation of water-in-oil emulsions, namely their tendency to separate into the water and oil constituents in the presence of contaminants in the fluid. Some separation may also occur in a clean fluid on standing in the reservoir although this is not normally significant. The separated oil-rich layer at the top is soon dispersed and re-mixed by agitation once the system is started up and flow is drawn from the reservoir.

The main attraction of water-in-oil emulsions is that they are the least costly of the fire-resistant fluids and thus are normally a first choice for industrial applications where a fire-resistant fluid is required. Limitations on lubricity and susceptibility to shear breakdown are least significant in such systems, since components generally work with generous clearances and very high pressures are not involved.

The fire-resistant quality is imparted by the water content, and is thus directly proportional to that content – i.e. the higher the water content the more fire-resistant the fluid. Fire resistance comes from the water being transformed into steam at high temperatures, imparting a snuffing action on any flame and carrying away the oil content.

TABLE 8.2

REPLACEMENT OF MINERAL OIL WITH FIRE-RESISTANT FLUID

<table>
<tr><th></th><th>Water/Glycol</th><th>Phosphate-Ester</th></tr>
<tr><td>1. Drain off existing fluid</td><td colspan="2">Particular attention to reservoir, lines, etc., where fluid may be trapped.</td></tr>
<tr><td>2. Clean</td><td colspan="2">Clean residue sludge, if necessary.</td></tr>
<tr><td>3. Painted surfaces</td><td colspan="2">Check for Compatibility</td></tr>
<tr><td></td><td>Some softening action on normal paints</td><td>Resistant paints must be used.</td></tr>
<tr><td>4. Seals, gaskets and elastomers</td><td colspan="2">Check for Compatibility</td></tr>
<tr><td></td><td>Probably compatible, except cork.</td><td>Replace to fluid maker's specification or recommendations.</td></tr>
<tr><td>5. Filter removed</td><td>Remove or disconnect</td><td>Remove element.</td></tr>
<tr><td>6. Flush</td><td colspan="2">Flush system with suitable flushing charge and drain whilst still warm.</td></tr>
<tr><td>7. Filter replaced</td><td colspan="2">Install suitable filter or new filament.</td></tr>
<tr><td></td><td>Active earth filament must not be used.</td><td></td></tr>
<tr><td>8. Examine pump, etc.</td><td colspan="2">Replace badly worn or damaged parts.</td></tr>
<tr><td>9. Fill with new fluid</td><td colspan="2">Make up to correct level.</td></tr>
<tr><td>10. Start up</td><td>Operate at reduced pressure initially and bring up to standard operating conditions.</td><td></td></tr>
<tr><td>11. Check filters</td><td colspan="2">Check twice a day for first week, once a day for next three weeks, then normal intervals.</td></tr>
<tr><td>12. Leaks</td><td colspan="2">Check and rectify (use drip pans to collect fluid, if necessary).</td></tr>
<tr><td>13. Residual oil</td><td>Compatible up to 5% max.</td><td></td></tr>
</table>

The fire-resistant properties will tend to decrease with age, due to evaporation of the water content. It is necessary to check the water content at regular intervals, and make good as necessary. To avoid excessive water evaporation the working temperature of the fluid should be kept as low as practical. A maximum working temperature of 160 degrees F (70 degrees C) can be utilised without excessive water loss, however, and so water-in-oil emulsions are more or less directly comparable with mineral oils in this respect. They have another similarity with oil fluids in that the same seals can normally be used, although cork and leather are to be avoided as seal materials.

It is an advantage with water-in-oil emulsions to protect the reservoir from heat and cold, as this will minimise variations in viscosity and vapour pressure. For the same reason lines between the reservoir and pump should be as short as possible, avoiding sharp bends; and the use of fine filters on the suction side of the pump should be avoided. A 10 micron filter on the pressure side is generally to be recommended, preferably with a metallic rather than a paper element.

Water-glycol Mixtures

Water-glycol mixtures consist of solutions of ethylene glycol or high molecular weight polyglycols, with corrosion inhibitors and load-carrying additives. The latter are necessary to compensate for the poor corrosion protection and poor lubricity, respectively, of the basic solution. Even with additives, however, the lubricating properties may be marginal for use with high pressure gear pumps or highly loaded rolling bearings. In general, however, water-glycol mixtures are capable of providing satisfactory lubrication, provided bearing loads are derated – e.g. see Table 8.3.

TABLE 8.3
COMPARISON OF FLUID LUBRICATING PROPERTIES

Fluid	Pump Life	General Lubricity	Suitability for Rolling Bearings
Water	Comparable with suitable construction	Very poor	Very poor
Mineral Oil	Standard	Very good to excellent	Very good
Water-in-oil Emulsions	Reasonable	Fair	Fair
Water-Glycol	Comparable at lower pressures	Fair	Poor
Phosphate Ester	Comparable	Good	Very good
Phosphate Ester Blends	Comparable	Very good	Very good
Chlorinated Hydrocarbons	Comparable	Very good	Very good
Silicones	Fair to comparable	Poor to fairly good	Poor to good

Water-glycol mixtures were chosen by the US Naval Research Laboratory in 1943 as the most promising basis for the development of fire-resistant hydraulic fluids, the principle reasons for the choice being the complete non-inflammability of such fluids (which have a 'snuffer' action), relatively low cost, good availability and without the disadvantages of high specific gravity and toxicity characteristic of many of the synthetic fluids. Such mixtures are generally compatible with conventional elastomers and standard seals, and to a limited extent with small residual amounts of oil fluids which may remain in a system on a changeover. Water-glycol solutions, with suitable additives, possess detergent properties, in fact, making them capable of holding in suspension any deposits formed in operation of the system until they are collected by the filters.

Water-glycol fluids were, however, subsequently abandoned by the U.S. Navy for their aircraft after some six years of trial, because of internal system corrosion problems which continued to occur. This was galvanic or electrolytic corrosion, initiated by the ability of the fluid to conduct small galvanic currents created by the use of dissimilar metals in a hydraulic system.

Considerable development has taken place with fluids of this type, particularly in the matter of thermal stability (elimination of separation at temperatures up to boiling point, and elimination of cloudiness and solidification at extremes of low temperature); and in anti-corrosion and wear properties. A range of grades of different viscosities can also be produced – eg typically from about 25 to 30 to 300 centistokes at normal ambient temperatures, depending largely on the water content. All such fluids are generally characterised by a high viscosity index of the order of 150 to 160 and thus show less variation of viscosity with temperature than most other types of fluids. A particular characteristic is usually a rapid fall in viscosity approaching 200–220 degrees F (circa 100 degrees C), but maximum working temperature is normally restricted to 150–165 degrees F (70 degrees C) in order to avoid excessive loss of water and of volatile additives by evaporation.

The shear stability of a water-glycol solution is also generally excellent – usually superior to that of a mineral oil fluid with viscosity index improvers, in fact.

Water content in typical water-glycol solutions may range between 35 and 60 per cent, although a more normal figure is an initial water content of around 35 per cent with a plus or minus 5 per cent tolerance against changes in use. Any variation outside this tolerance needs compensating by the addition of make-up fluid. Normally this will be water as a water-loss is more likely than a glycol-loss.

Since the water content of a particular fluid largely governs the viscosity, a viscosity reading can be taken as a direct measure of the water content at any stage of the fluid life. More specifically, a viscosity reading can be calibrated to indicate the water deficiency of a particular fluid, and thus the amount of make-up water required – e.g.see Fig 8.2. As with water-in-oil emulsions, any make-up water used must be pure, e.g. distilled water, de-ionised water, or boiler condensate water.

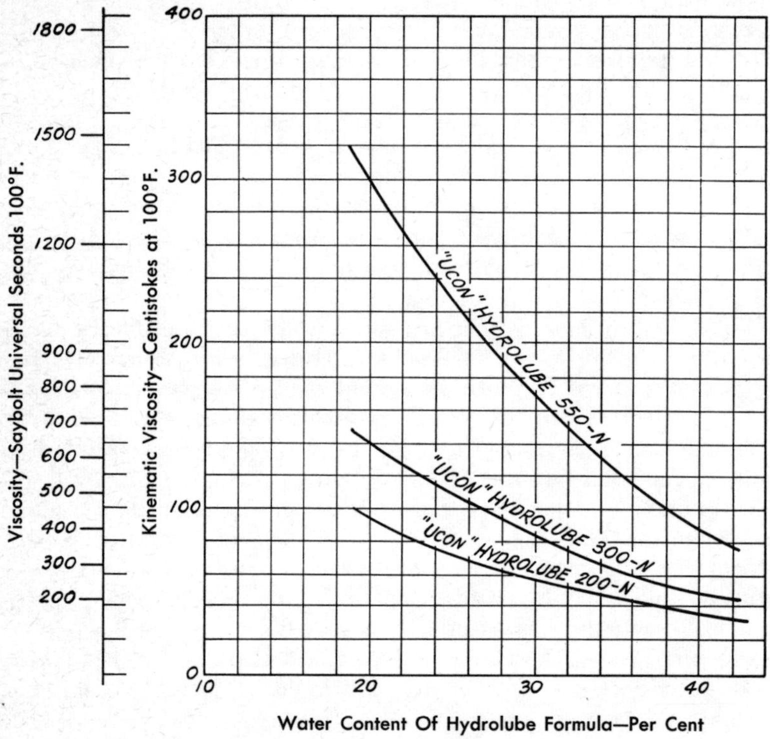

Fig. 8.2 Effect of water content on the viscosity of "UNCON" Hydrolubes.

In many cases the additive content can also be checked by a measurement of the alkalinity. Loss of alkalinity is an indication of loss or degradation of inhibitors; or, equally, a positive measurement or indication of alkalinity confirms that a suitable amount of inhibitor remains in the fluid. Such treatment cannot be taken as a general rule since the type, and proportions, of inhibitors can vary considerably with different fluids. Where the general method is applicable, however, this provides a simple method of checking the state of the inhibitor(s) in the fluid, consistent with specific alkalinity measurement data (e.g.neutralisation test on a sample) stated by the manufacturers of that fluid.

Although water-containing fluids, water-glycol solutions (with inhibitors) are not seriously corrosive to metals. They do not attack steel, copper, brass or aluminium, ie the usual materials of construction for hydraulic components, but can promote electrolytic action at bi-metal couples in these materials. Cadmium, magnesium and zinc are all slightly attacked by water-glycol solutions, so the use of magnesium alloys or cadmium or zinc plating should be avoided.

An important point in connection with the corrosion resistance of water-glycol solutions is that the inhibitors used are effective only in the liquid phase, ie vapourised fluid may be potentially corrosive. For this reason the insides of reservoirs should be painted with an approved paint scheme.

Glycol-water solutions are not compatible with other types of fluids. On a changeover an existing system should therefore be flushed out thoroughly. Existing seals and elastomers should be compatible, if originally used with mineral oil fluid, although their suitability should be checked as a commonsense precaution.

The bulk modulus of water-glycol solutions is high, the compressibility of such fluids being less than that of water or water-in-oil emulsions – see Fig 8.3.

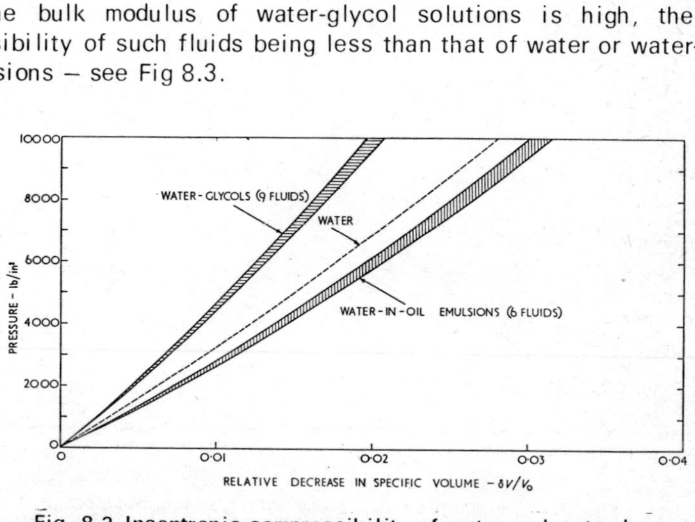

Fig. 8.3 Insentropic compressibility of water and water lace fluids at 20°C (NEL Measurements.)

The vapour pressure of water-glycol solutions is higher than that of comparable mineral oils, which emphasises the need for care in the design of the suction side of the system. Restriction must be avoided on the suction side and the pump placed as low as possible in relation to the reservoir level in order to ensure satisfactory suction entry without cavitation of 'flashing'.

The pour point of all water-glycol solutions is low, the depression of the freezing point being directly related to the proportion of glycol (eg as in anti-freeze mixtures). Thus water-glycol solutions remain fluid at relatively low temperatures.

Pour point is an important property since fire-resistant fluids tend to aggravate cold starting problems. If the viscosity is too great the fluid will be slow to reach working surfaces; high wear rates can occur on starting up the system. This applies regardless of how good a lubricant the fluid is, but the problem is obviously aggravated if the fluid is not a good lubricant. A satisfactory pour point is thus a necessary requirement of any hydraulic fluid.

Similarly, thickening of the fluid with resulting increase in viscous friction can affect the performance of the pump on the suction side. Also all fire-resistant fluids have higher specific gravities than oil fluids, which again tend to make the suction side of the system more critical in design and performance.

Phosphate Ester

Phosphate ester is a fully synthetic fluid which approaches most closely the desirable properties of a mineral oil fluid, namely good

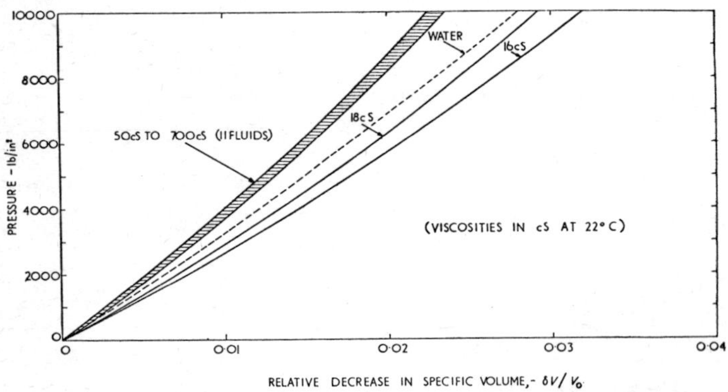

Fig. 8.4 Insentropic compressibility of phosphate ester fluids with viscosities ranging from 16 centistokes to 700 centistokes at 22 ^{0}C (NEL Measurements.)

TABLE 8.4
COMPATIBILITY OF FIRE-RESISTANT FLUIDS*

Type of Fluid	Corrosion Protestion Characteristics	Suitability or Standard Seals	Seal Materials Not Suitable	Special Seal Materials	Metals to be Avoided	Effect on Paint
Water-in oil Emulsion	Good	Generally satisfactory	Cork, leather, resin-bonded paper	None	Cadmium Zinc	None
Water-glycol	Fair	Generally satisf actory	Cork	None	Cadmium Zinc	Softens
Phosphate Ester	Fair	Incompatible	Most elastomers except Butyl & EPM	Ethylene - propylene & Butyl	Aluminium bearings	Stripping action
Phosphate Ester Blends	Fair to good	Incompatible	Most elastomers except Butyl & EPM	Ethylene-propylene & Butyl	Aluminium bearings	Stripping action
Chlorinated Hydrocarbon	Fair	Incompatible	Most elastomers except Butyl & EPM	Ethylene-propylene	Aluminium bearings	Stripping action
Silicone	Fair	Generally satisfactory up to 300°F	—	Viton (For high temperatures)	Several, but dependent on temperature	None

lubricating properties, good anti-corrosion properties and good chemical stability. It is also a non-inflammable fluid although under extended heating some decomposition will take place and the resulting vapour products may burn. It is therefore correctly classified as a fire-resistant fluid. Thermal stability is excellent up to more moderate temperatures and the maximum working temperature for the fluid is substantially higher than that of mineral oils, e.g. about 300 degrees F (150 degrees C). Compressibility characteristics are shown in Fig 8.4.

The main limitations of phosphate ester are: (i) high cost; (ii) incompatibility with most conventional elastomers, calling for the use of special seal materials; (iii) a low viscosity index; (iv) high specific gravity; and (v) a stripping action on conventional paint finishes.

Additives may be used with phosphate ester, notable viscosity index improvers, although these will reduce the shear stability of the fluid – i.e. may be associated with a temporary or permanent loss of viscosity at high shear rates. Lubrication properties are generally excellent, and in some functions better than mineral oils (especially under lighter loads); although some reduction in bearing loading is generally advisable where the loading is severe (e.g. in rolling bearings and vane and gear pumps). The fluid is non-corrosive to steel, copper, brass and aluminium, but may show slight attack on cadmium, magnesium and zinc. Corrosion inhibitors may be added to impart extra rust protection in steel systems.

The high cost of phosphate ester fluids normally limits their industrial applications to small or moderately sized systems, or to those where the higher permissible working temperature of the fluid is advantageous (e.g. systems working continuously at 140 degrees F or above). The higher specific gravity of the fluid is also an advantage for use in hydrokinetic systems, since this will result in a proportionate increase in torque capacity of a unit compared with mineral oils. This is even more marked in the case of chlorinated hydrocarbons and phosphate ester blends, although cost is usually the deciding factor.

For aircraft hydraulic systems cost is largely inconsequential and phosphate esters have become virtually the standard fire-resistant fluid for (mandatory) civil aircraft systems.

The vapour pressure of phosphate ester fluids is low, but the suction side of the system again needs careful design to take into account the higher specific gravity and, particularly, the low viscosity index. Ideally the suction side of the system should be protected against low temperatures.

Initially sealing presented severe problems with phosphate ester fluids since only a relatively few elastomers proved compatible. The problem is also individual to the exact blend or fluid used, as well as the service conditions of the system. Butyl rubber may be satisfactory with phosphate ester (but less so with blends containing aromatic hydrocarbons). Ethylene propylene, or EPDM, are both generally suitable for use with phosphate esters and are now normally the first choice material. Fluorocarbon rubbers and PTFE are alternatives for more specialised applications.

Because of the stripping action, internal painting should be avoided as far as possible. Damage to external paintwork may also occur at 'spillage' areas. Certain paints are, however, compatible and so if painting is essential, a paint manufacturer's recommendations should be sought and followed.

Phosphate ester fluids are completely incompatible with other types of fluids and so on a changeover, complete and thorough flushing of the system is necessary. As a general rule, all seals and elastomers must also be changed, although Viton is one elastomer which is compatible with both mineral oils and phosphate esters.

Chlorinated Hydrocarbons

This category of synthetic fluids contain up to 40 or 50 per cent by weight of chlorine (as opposed to a typical phosphorus content of 8 per cent in phosphate esters), have an even higher specific gravity (around 1.4 to 1.45), and are even more expensive. Properties as functional fluids are similar to phosphate esters, but with certain improvements.

Because of their high cost chlorinated hydrocarbons are seldom used on their own as hydraulic fluids, but mixed with phosphate ester in varying proportions to produce blends of different specific gravities and viscosities. Viscosity index improvers are normally incorporated in such blends to improve the viscosity-temperature characteristics, chlorinated hydrocarbons having an even lower viscosity index than phosphate esters. Similar precautions apply as regards the suction side of the system, and as regards seal compatibility.

One specific property of phosphate esters, chlorinated hydrocarbons, and blends, is that all such fluids have a high bulk modulus – rather better than water and substantially better than mineral oils. This may favour their selection for high pressure systems where rigidity is important, regardless of whether or not fire-resistance is significant.

Silicone Fluids

Silicone fluids are essentially non-inflammable and have an extremely high maximum service temperature of the order of 600 to 700 degrees F (depending on formulation). Because of their extremely high cost they are, however, more properly regarded as high-temperature rather than fire-resistant fluids, i.e. would normally only be selected for systems required to operate at high working temperatures. They are also suitable for working over a wide range of temperatures since the viscosity index of such fluids is high.

Straightforward silicone fluids, in general, suffer from poor lubricity and anti-wear properties. Such limitations have largely been overcome by development of these fluids, although certain limitations may remain on the question of compatibility. Conventional elastomers are generally compatible with silicones and modified silicones up to the maximum service temperature of the material, but some leaching out of plasticiser may occur. Since silicones are normally used for services with working temperatures above the maximum limit for conventional elastomers, choice of seal material becomes strictly limited, only Viton being a generally suitable material for use up to 500 degrees F (260 degrees C). For even higher working temperatures metallic (reed) seals may have to be considered.

Modified silicones also tend to attack metals, especially at elevated temperatures. Thus for a high-temperature, high-pressure system, pipework and components would normally be specified in stainless steel, preferably with welded connections, further adding to the cost of the system.

High Temperature Fluids

Although there is the possibility of using mineral oil fluids for systems working in high ambient temperatures by providing adequate intercooling, or at even higher temperatures by such artifices as inerting the system by flooding through with nitrogen or a similar inert gas, there are limits at which available organic fluids are unsuitable.

Oxidation and corrosion stability problems are aggravated with increasing temperature, and thus again the demand becomes increasingly difficult, as no insoluble materials or sludge can be tolerated in a high duty system. And any loss of viscosity in an otherwise satisfactory fluid must remain within acceptable limits.

Hydrolytic stability is another important feature, since fluid both in storage and in the working system may become contaminated with water. Some otherwise satisfactory fluids may break down under such conditions, others become excessively acid, oxidise and promote corrosive attack, even to the extent of producing seizure in some moving parts or depositing insoluble matter through the system to clog orifices, etc.

All other factors being acceptable, thermal stability will then ultimately determine the maximum operating temperature of the fluid.

There will be some temperature at which the fluid has dissociated or deteriorated to a degree where its properties are no longer acceptable, such as increasing acidity or loss of viscosity, etc. The volatility index is also significant as affecting pump performance — see Fig 8.5.

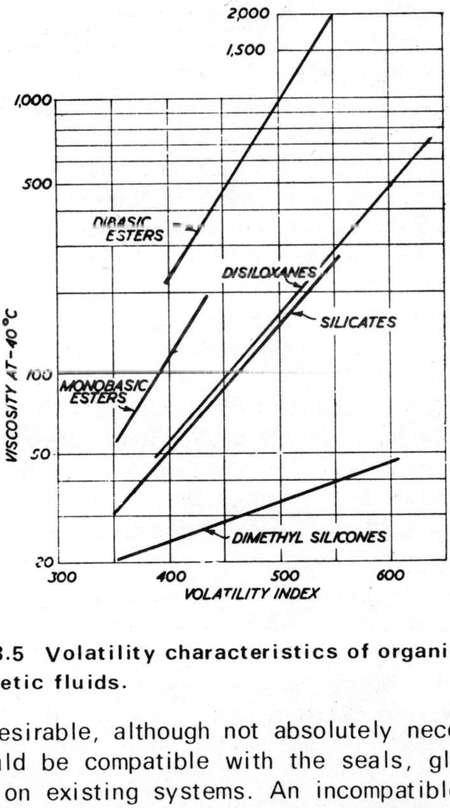

Fig. 8.5 Volatility characteristics of organic and synthetic fluids.

It is also desirable, although not absolutely necessary, that any new fluid should be compatible with the seals, glands and hoses currently used on existing systems. An incompatible system is not necessarily a disadvantage if designed as such from the start, except for the possibility of accidentally topping up with an incompatible fluid. But to change the seals, etc, on an existing system is generally an expensive project in terms of man-hours. In a compatible system merely flushing out with the new fluid is sufficient before complete replacement. It is an unfortunate fact that where two incompatible hydraulic fluids are accidentally mixed it is nearly always the viscosity-improver additives which precipitate out in the form of gums.

Synthetic rubber seals are more or less standard throughout hydraulic systems. Most hydraulic fluids have some swelling action on synthetic rubber, and, indeed, a certain amount of swelling is desirable as promoting a better seal. The question of degree of swelling acceptable is rather debatable, a generally acceptable limit being of the order of 25 per cent swell.

Foaming is a characteristic frequently advanced as an important criterion, and although excessive foaming can give trouble under cycling conditions in a hydraulic system the solution lies largely in the design of the system itself. Foaming should never be a problem in completely closed systems.

When dealing with high-temperature operation, the question of specific heat and thermal conductivity becomes increasingly important as governing the size of the heat exchanger necessary to remove excess heat from the hydraulic system. Removal of such heat is, in fact, becoming a serious problem with supersonic aircraft installations. Thus fluids with high specific heat and high thermal conductivity are to be preferred.

Other factors to be taken into account are cost and availability of the fluid, its density as affecting both the weight of fluid and hydraulic ram effect, compressibility in affecting response time and static rigidity of the system, toxicity for general safety in handling, and its possible effects on paints and finishes and electrical insulation. Some organic esters, particularly the phosphate esters and di-esters, have a particularly strong solvent action on such materials, and although resistant materials have been produced these are not commonly available.

Consideration of viscosity characteristics and volatility largely rules out the dibasic acid esters, polyhydric alcohol esters, orthosilicates and derivatives of polyglycols. The silicates, in general, are prone to hydrolytic instability and thermal instability. Silicones, in general, have very poor lubricity. Silicones and silicate esters, on the other hand, possess excellent physical properties over the range of temperature required.

Thus development of practical high-temperature fluids has, largely, concentrated on taking a basic synthetic with such starting properties and reworking or developing it to enhance its original limitations. In this respect, modified silicones appear the most practical solution to date and are, in fact, the only types readily available commercially. Hydrolysis can be a very real problem, but despite their apparent limitations at the upper temperature limit, silicate-base fluids have shown excellent stability at 200 deg F and could certainly be worked satisfactorily at service temperatures considerably in excess of this figure. There is also a noticeable trend to develop sealed 'dry' systems for high-temperature hydraulics which could provide a complete answer in this respect.

SPECIFICATIONS OF PROPRIETARY FLUIDS TO COME

Liquid Metals as Hydraulic Fluids

Basically, interest in the possibilities of liquid metals as a hydraulic fluid has been triggered by the environmental conditions of hypersonic aircraft and the desirability of retaining hydraulic flight control actuating systems with fluid operating temperatures of 1 000 deg F or above.

This is well beyond the operating range of even the best of current high temperature fluids (silicones), or even silanes or bi-esters (experiemntal high temperature fluids). The alternatives to liquid metal fluids are essentially compromises – eg the use of 'conventional' fluids but with the system operating temperature suitably reduced by a cooling system; or alternative actuating systems. As regards the former, cooling is, in fact, current practice on a number of hypersonic aircraft. Thus the XB-70 Mach 3 aircraft utilised hydraulic system cooling to maintain the fluid temperature below 425 deg F, the fuel carried by the aircraft being used as a heat sink. This is certainly a practical method, although implying further complication and weight penalty. Thus the XB-70 had to carry some two tons of water purely for additional cooling when the fuel level was low. The rate of heat transfer necessary reached as high as 20 000 btu *per minute.*

The other alternatives – high pressure pneumatics and all-mechanical systems – are also under developmental investigation as possible high temperature systems, but are less attractive for the basic functional requirements of reliable high-level power transmission. Pneumatics in particular, also suffer from lack of precision. To add precision with hydraulic damping or hydraulic checking simply re-introduces the fluid cooling problem, although not necessarily on the same scale. Thus the desirability of utilising a hydraulic system remains unaltered.

The problem is not necessarily one associated only with the development of hypersonic aircraft, missiles and spacecraft. A high temperature hydraulic system has obvious industrial applications for use in proximity to high heat sources. The immediate limitation of a liquid metal system is then its 'freezing' temperature, when it may not be practical to use a metal whose melting point is higher than room temperature unless the whole of the system is continually maintained at a higher temperature. The same is also true of fusible salts which have been suggested as alternatives to liquid metals.

Liquid metals, on the other hand, are undoubtedly the most 'logical' of the high temperature fluids in that they have a ready ability to function as hydraulic fluids at 1 000 deg F and above; with the further advantage that they will remain stable under conditions of high nuclear radiation. This again can have industrial significance. The operating temperature range over which the metal remains liquid can also be extended by the choice of suitable alloys. Here the range covered is, in fact, likely to be more suitable for specialised industrial applications than aircraft, where conventional requirements require that the operating range be extended down to –65 deg F.

Probably the most promising metal to date in this respect is a eutectic alloy of 23 per cent sodium and 77 per cent potassium, designated NaK-77. This alloy has a melting point as low as 10 deg F and a boiling point of about 1 450 deg F at atmospheric pressure. The alloy undergoes no degradation at all with temperature or time over the whole of this range, and in this respect alone offers unique properties as a hydraulic fluid.

On the other hand, it is violently reactive with oxygen or moisture so that special precautions are necessary to prevent fluid contamination; has poor lubricating properties; and needs special designs of seals and pumps to handle, as well as specialised valves and actuators.

Considering the 'fluid' properties first, as well as being a very stable fluid NaK-77 is also comparable in density to conventional hydraulic fluids and a viscosity somewhat lower than that of water – Fig 8.6, The viscosity is, of course, dependent on temperature and ranges from 1.8 lb/hr-ft in absolute units at its melting point (10 deg F) to around 0.4 lb/hr-ft at 1 000 – 1 200 deg F. Vapour pressure ranges from about 1 mm of mercury at 600 deg F to 100 mm of Hg at 1 000 deg F and 1 000 mm of Hg at 1 500 deg F – Fig 8.7. Thermal and electrical conductivity is high, as would be expected of a metal, but no specific figures have been made available.

Electromagnetic, vane and centrifugal pumps have all proved workable for handling the liquid metal fluid, although the former have limited power capabilities. The fact that the fluid is electrically conductive, however, makes electromagnetic pumping attractive and the principle has been embodied in some specialised designs of servo valves. Mechanical pumps are the logical choice for pressurisation and the centrifugal type would appear superior to the vane although no conclusive results have yet been reached as to the best pump configuration. Most pumps which have been evaluated have been modifications or redesign of standard available aircraft hydraulic pumps with high temperature potential, the main modifications being concerned with the bearings and seals, and compatibility of metallic elements.

Suitable materials appear to be molybdenum and certain nickel-chromium alloys, with tungsten and tungsten carbide bearings. Alkali

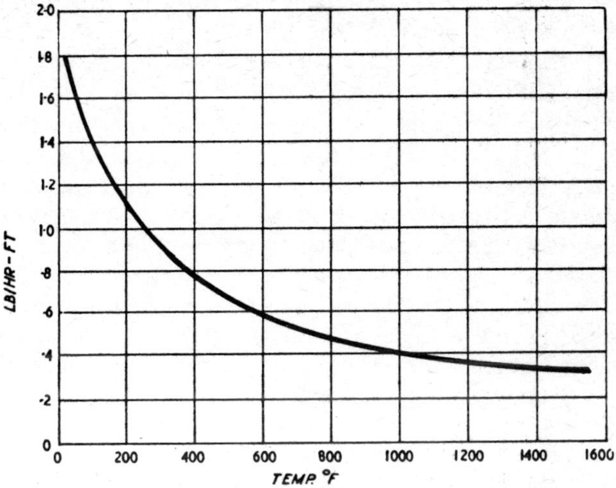

Fig. 8.6

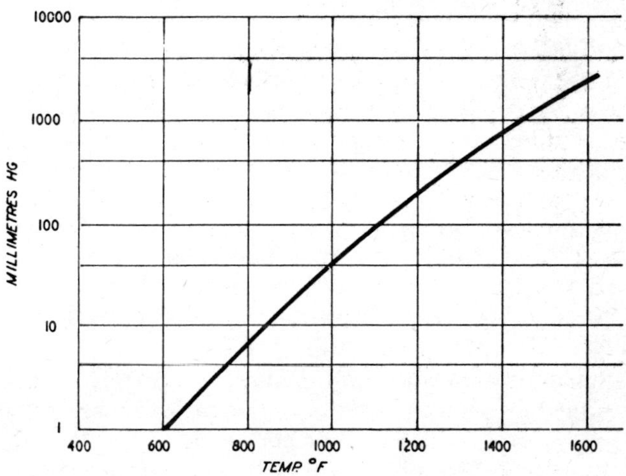

Fig. 8.7

metal films are extremely poor lubricants and bearing design tends to be critical, if seizure is to be avoided. On the other hand pressurised hydrodynamic bearings lubricated by the pumped liquid metal with bellows type shaft seals do appear to be capable of giving satisfactory service. The latter are all-metallic rotating face seals in metals capable of continuous operation at temperatures in excess of 1 000 deg F and speeds of up to 35 000 rpm. Lubrication at best, however, tends to be marginal, this particular feature being the most critical of the whole pump design.

Valves do not appear to present any particular problem, other than compatibility of materials and the accommodation of thermal expansion at elevated working temperatures (eg with low expansion metals) One particular design of servo valve developed by the General Electric Co, however, utilises a small electromagnetic pump as a combination transducer and pilot stage in a two-stage configuration. The first stage comprises a pumping section and two bellows filled with NaK-77 and hermetically sealed, with its response controlling two poppet valves in the second stage. The response is in the form of a differential pressure between the two first stage bellows produced by the magnetic pump responding to input (signal) current both as regards quantity of signal value and direction. The result is to maintain a fixed ratio of second stage output pressure to control pressure within the flow capabilities of the valve (up to 10 gpm at 2 000 psi).

The same company have also produced a specialised design of actuator in 422 stainless steel (chosen for its low coefficient of thermal expansion) with a carbide bearing and cylinder liner. The latter are installed by shrink fitting at a temperature differential of 930 deg F and dimensioned to ensure that interference fit is retained to 1 300 deg F. Piston and rod are flame plated with tungsten carbide, thus providing two different hard surfaces in rubbing contact to minimise friction and wear. Piston sealing is by conventional metallic piston rings, again flame plated in tungsten carbide. The rod seal comprises a carbide bearing acting as an inner capillary seal together with an external metal reed-type seal. This particular linear actuator was designed with a 2 in bore and 4½ in stroke for operating at system pressures up to 3 000 psi and temperatures up to 1 200 deg F, specifically with NaK-77 liquid metal fluid.

Certain conventional all-metallic static seals have proved satisfactory for limited service with liquid metal systems, although the evolution of special seal forms is necessary for extended service, together with the elimination of seals altogether where possible. Thus where the fitting of a seal may represent a simple and convenient solution with conventional hydraulics, each sealing point on a high temperature system represents an exacting problem for solution.

Although metallic seals were used successfully in conventional hydraulic systems as far back as 1936, one of the major problems which has hitherto retarded their further development is lack of knowledge of the frictional characteristics of metals in rubbing contact at elevated temperatures. Thus metals which behave satisfactorily at low to moderate temperatures may show excessive wear or early failure under rubbing or sliding conditions at higher temperatures. This problem is, of course, specific to dynamic seals. Static seal requirements for elevated service temperatures and pressures up to 3 000 psi can generally be met satisfactorily with metallic O-rings or similar devices, provided they are installed properly.

The metallic reed seal in its basic form consists normally of a simple conical washer shape with or without a lip and anchored at the apex end in a suitable groove or housing. This has proved to be a successful type of dynamic seal, within the limitations of the material stresses involved, and wear which may occur at the tip. It is pressure energised as well as being preloaded to contain static pressure and can accommodate rotary as well as linear motions. Reed seals of this type have been employed on a somewhat limited scale in conventional hydraulic systems, but the reed seal used with the liquid metal actuator described above is a new design.

Experience to date with liquid-metal hydraulic systems would appear to indicate that, apart from sealing problems and the lubrication of sliding parts (particularly in pumps), NaK-77 behaves in an entirely predictable manner and with characteristics similar to that of any normal fluid of similar viscosity. The sealing problem is, of course, aggravated by the low viscosity of NaK-77, just as it is more difficult to provide sealing and lubrication with a high pressure water system compared with an oil-hydraulic system. The question of 'additives' is entirely ruled out since the metallic fluid may be used, and remain, in a pure metallic state as all of the alkali metals are chemically active in the presence of oxygen-containing contaminants or moisture.

At least, however, investigations carried out to date have established that the liquid-metal hydraulic system is certainly feasible and workable at temperatures up to or even slightly exceeding 1 200 deg F. The most promising fluid – NaK-77 – also remains liquid down to a temperature of 10 deg F so that the system remains fluid at all normal atmospheric temperatures. The cost, however, will be extremely high and prohibitive for all but specialised applications where hydraulic power at temperatures of the order of 1 000 deg F shows advantages over alternative high pressure pneumatic or mechanical systems. Each component in the system needs to be specially designed, probably with extensive use of such high cost materials as molybdenum and carbides, and there is still a considerable amount of development work to be undertaken to turn test system experience into practical and reliable components with an acceptable service life.

PROPRIETARY FIRE-RESISTANT FLUIDS

CASTROL FIRE RESISTANT HYDRAULIC FLUIDS – SUMMARY OF PROPERTIES

	Hyspin FR	Hyspin FR Light	Hyspin AF1	Hyspin AF2	Hyspin AF3	Hydromul 68	Hydromul 100
General type	Phosphate ester.		Water glycol solution			Water in oil emulsion	
S.G at 60/60 °F	1.145	1.120	1.093	1.073	1.062	0.924	0.925
pH value	not applicable		10.4	10.4	10.4	not applicable	
Viscosity cS							
at 70 °F			107	46	26	150	220
at 100 °F	48	19.2	53.6	23.9	14	67	108
at 140 °F			26.8	12.0	7.4	31.5	46.5
at 210 °F	4.8	3.5	10.6	5.1	3.4	—	—

†V.I.	0	0	215	135	130	142	142
Pour point °F	−5	−30	−50	−15	5	−45	15
Spontaneous ignition temp. °F	1150	1150	887	900	900	not applicable	
Effect on paint	Severe but selective		Severe but selective			No more than mineral oil	
Fire resistance							
Wick test	pass	pass	not ignited			pass	pass
Spray ignition test	pass	pass	not ignited			pass	pass
Vapour pressure mm Hg	0.1 @ 400°F	0.1 @ 400°F	17 @ 70°F	18 @ 70°F	19 @ 70°F	20 @ 70°F	20 @ 70°F
Specific heat at 60°F	0.40	0.38	0.71	0.79	0.83	0.67	0.67

†$V.I._E$ except for Hyspin FR and FR Light.
*At room temperature.

LORCO WATER/GLYCOL FLUIDS

		Lorco WG 'L'	Lorco WG 'MH'
Kinematic Viscosity	@ 100 deg F	33.5 cs	59.0 cs
	@ 210 deg F	7.3 cs	12.6 cs
Redwood Viscosity	@ 70 deg F	260 in	450 in
	@ 100 deg F	137 in	240 in
	@ 140 deg F	73 in	125 in
	@ 200 deg F	47.5 in	62.5 in
Viscosity Index		165	155
Specific Gravity	@ 60 deg F	1.079	1.080
pH		10.0	10.0
Pour Point		−40 deg F	−25 deg F
Water Content		45%	45%

LORCO WATER IN OIL FLUID

Typical Specification of LORCO INVERT 35

Specific Gravity	0.930 @ 60 deg F
Colour	Blue
Pour Point	+ 15 deg F
Redwood Viscosity	850/900'' @ 70 deg F
	160/165'' @ 140 deg F
Viscosity Index	140
Neutralization Value	0.5 mgms KOH/gm
Water Content	40%
Specific Heat	0.68
Isentropic Secant Compressibility	3.08×10^{-6} in^2/lb

LORCO PHOSPHATE ESTER FIRE RESISTANT FLUIDS

Grade		PE/B	PE/L	PE/MH	PE/H	PE/XH
Kinematic Viscosity in Centistokes	210°F	3.3	4.2	5.1	6.0	6.9
	100°F	17.5	29.0	43.0	75.0	120.0
	10°F	1,000	4,000	20,000	–	–
Viscosity Index		+32	<0	<0	<0	<0
Temperature in °F for Viscosity of	850 c.St	5	32	41	59	70
	3.5 c.St	205	230	250	260	275
Pour Point in °F		–30	–15	–5	+15	+20
Maximum Specific Gravity at 60°F		1.210	1.175	1.160	1.160	1.160
Flash Point in °F COC		475	505	480	505	510
Auto Ignition Temperature in °F		1,238	1,229	1,139	1,148	1,157
Acid Value in mg/KOH/gm maximum		0.1	0.1	0.1	0.1	0.1
Water Content in % maximum		Nil	Nil	Nil	Nil	Nil
Boiling Point in °C at 760 mm/Hg		412	413	417	414	407
Coefficient Thermal Expansion in cc/cc°C		6.8×10^{-4}	6.9×10^{-4}	6.9×10^{-4}	6.9×10^{-4}	6.9×10^{-4}
Specific Heat in Cal/gm/°C at 25°C		0.375	0.360	0.378	0.385	0.388
Thermal Conductivity in Cal/sec.cm²/°C		Average value 0.0003				
Vapour Pressure in mm/Hg at 220°C		5.0	4.5	5.5	3.2	2.5
Surface Tension in dynes/cm²		44.96	41.95	39.73	38.98	40.71

MOBIL NYVAC FLUIDS

Mobil Nyvac No.20 and No.30 are two orange/red coloured, fire resistant, water based fluids of the glycol type recommended for use as the hydraulic media in machines which are operated in close association with a source of ignition. Due to their water content, they are not recommended for use at bulk oil temperatures above 150 deg F. Typical applications include die-casting machines, press welders, furnace door openers, combustion regulators, ladle tilting mechanisms and glass drawing machinery.

Mobil Nyvac Inhibitor WB is a special alkalinity control concentrate which is added – either to the fluid – or to the make-up water to restore Mobil Nyvac fluids to their original viscosity after water evaporation during service has caused additive material depletion.

Typical physical characteristics of the Mobil Nyvac fluids are:-

	Mobil Nyvac	
	No.20	No.30
Specific Gravity	1.08	1.08
Pour Point °F (max)	–25	–20
Viscosity at 100 °F	42.5 Centistokes	64.5 Centistokes
	200 SSU	300 SSU
	178 Redwood No.1	267 Redwood No.1
Viscosity Index	160	160
Colour	Orange/red	Orange/red
Water Content	45%	45%

MOBIL PYROGARD 160

Mobil Pyrogard 160 is a straight chlorinated hydrocarbon type fire resistant fluid recommended for hydraulic systems and as a lubricant for conditions where fire hazards exist.

It contains a rust inhibitor, a combined anti-oxidant and anti-wear additive, and a defoamant. Mobil Pyrogard 160 will not shear down to a reduced viscosity regardless of the length or severity of service, and is completely miscible with Pyrogard 210, the phosphate ester chlorinated hydrocarbon type material which it supersedes.

Fluid Type	Chlorinated Hydrocarbon
Specific Gravity at 60°F	1.42
Viscosity at 100°F	31.9 Centistokes
	150 SSU
	132 Redwood No.1 Seconds
Viscosity at 210°F	2.85 Centistokes
	35.8 SSU
	33 Redwood No.1 Seconds
Colour	Blue
Pour Point	30°F (max)
Flash Point C.O.C.	375°F (min)

MOBIL PYROGARD D

Pyrograd D is a ready mixed off-white coloured water-in-oil emulsion recommended for use as a fire resistant hydraulic fluid. It is suitable for use in all types of hydraulic pumps except radial piston designs which normally operate on fluids having a higher viscosity. The use of Pyrograd D in this type of pump would result in lower volumetric efficiency and increased wear,

Specific Gravity	0.935
Pour Point	Minus 20° F Max
Viscosity at 100°F	
c/s	86-108
S.S.U.	400-500
Redwood No.1	350-441
Water content	40% by volume

HOUGHTO-SAFE 5046
(Water-in-Oil Emulsion)

Redwood No.1 Viscosity @ 100°F	420 secs
Viscosity Index	120
Specific Gravity @ 60°F	0.956
Flash Point	Nil
Pour Point	−15°F
Water Content	40%
Colour	Off White

HOUGHTO-SAFE 1010 SERIES
(Phosphate Ester Fluids)

	Houghto-Safe 1010	Houghto-Safe 1120	Houghto-Safe 1055
Specific Gravity @ 60°F	1.208	1.145	1.150
Redwood No.1 Viscosity			
@ 70°F	168 secs	600 secs	–
@ 100°F	77 secs	183 secs	470 secs
@ 140°F	46 secs	70 secs	108 secs
Pour Point	– 30°F	– 5°F	20°F

HOUGHTO-SAFE 620
(Water-Glycol)

Redwood No.1 Viscosity	
@ 0°F	4500 secs
@ 70°F	350
@ 100°F	175
@ 140°F	83
Viscosity Index	140
Pour Point	−60°F
pH	8 to 10
Specific Gravity @ 60°F	1.055
Flash Point (ASTM)	None
Fire Point (ASTM)	None
Colour	Light Amber
Specific Heat	0.8
Thermal Conductivity	0.25 BTU per hr. per sq ft per °F per ft.
Maximum operating temp.	140°F

LORCO LOW TEMPERATURE HYDRAULIC OILS

(Naphthenic Base)

Method of measurement	Property (Grade / Viscosity Number)		AT 21	BT 24	NT 26	NT 32	NT 35
IP71/62	Kinematic Viscosity (Centistokes)	210°F	2.9	3.96	5.1	8.3	10.4
		100°F	13.6	22.8	36.0	87.0	130
IP70/62	Redwood Viscosity (Seconds)	0°F	1350	3750	10 500	60 600	155 000
		70°F	110	205	360	1 000	1 750
		100°F	64	95	147	340	520
		140°F	43.5	54	67	120	165
		200°F	34.5	37.5	41	51	59
	Viscosity Index		55	58	70	65	53
IP15/60B	Pour Point	0°F	−40°F	−25°F	−25°F	−25°F	−25°F
	Specific Gravity	60°F	0.875	0.885	0.890	0.900	0.908
IP34/58 / IP35/42	Flash Point °F	Closed	335 310	395 370	4[illegible] 390	[illegible] 405	445 430
		Open	335	395	410	420	445
IP2/61/A	Aniline Point	°C	73	86	90	92	93

TYPICAL PROPERTIES OF LORCO HIGH V.I. HYDRAULIC OILS

			FVT25	FVT32
IP71/62	Viscosity (Centistokes)	210°F 100°F	3.75 15.2	8.75 48.0
IP70/62	Viscosity Redwood (Seconds)	0°F 70°F 100°F 140°F 200°F	850 112.5 69 47.5 36.5	4 500 395 195 94 52
	Viscosity Index		155	145
IPk5/60B	Pour Point	°F	–40°F	–25°F
	Specific Gravity	60°F	0.875	0.885
IP34/58 IP35/42	Flash Point °F	Closed Open	310 325	400 420

EVCO HYDRAULIC OILS (Edgar Vaughan)

			Evco Light	Evco Medium	Evco Heavy	Evco Extra Heavy
Redwood No.1 Viscosity @	70°F	70°F	120 secs	331 secs	539 secs	687 secs
	100°F		68	144	219	269
	140°F		46	69	90	107
	200°F		35	42	47	51
Kinematic Viscosity @	70°F		29.0 cs	82 cs	132.5 cs	170 cs
	100°F		14.7	35.1	53.0	66.0
	140°F		7.5	15.3	21.4	26.1
	200°F		3.58	6.2	8.1	9.4
Viscosity Index			100	104	100	100
Specific Gravity @	60°F		0.875	0.879	0.879	0.881
Flash Point (0)	Pensky-Marten		320°F	350°F	360°F	370°F
Pour Point			−40°F	−22°F	−11°F	−4°F

HYDRODRIVE M.I.H. SERIES

Typical Physical Properties	Hydrodrive M.I.H.Light	Hydrodrive M.I.H.10	Hydrodrive M.I.H.20
Specific Gravity @ 60°F	0.880	0.882	0.887
Flash Point (0)	410°F	410°F	420°F
Redwood No.1 Viscosity			
@ 70°F	315 secs	445 secs	680 secs
100°F	135 secs	186 secs	270 secs
140°F	67 secs	83 secs	106 secs
Viscosity Index	95	95	95
Pour Point	5°F	5°F	10°F

HYDRODRIVE H.P. SERIES

	Hydrodrive H.P. 150	Hydrodrive H.P.200	Hydrodrive H.P.300
Specific Gravity @ 60°F	0.866	0.880	0.879
Redwood No.1 Viscosity			
(in Seconds) @ 70°F	315	440	600
100°F	140	185	250
140°F	68	80	105
200°F	42	44.5	52.5
Kinematic Viscosity @ 100°F	34 cs	46 cs	62 cs
Viscosity Index	100 min	100 min	100 min
Flash Point (Open)	395°F	415°F	420°F
Pour Point	5°F	5°F	10°F

9 HYDRAULIC SEALS

Pressure energised seals are the normal choice for hydraulic applications, using seal sections and seal materials which deform under pressure. Compression packings may, however, still be used for rod seals on large machines with reciprocating motions, particularly in water-hydraulic systems. Because high system pressures permit the use of smaller components single seal rings may be adequate, considerably simplifying seal selection and installation. For heavy duty applications, particularly in larger sizes of rams, seal sets are generally preferred. Here proprietary seal ring sets have largely superceded the traditional U and V-section rings for gland packings, because of the superior performance offered by these special sections.

The general suitability of various types of seals for pressure rating is summarised in Table 9.1, whilst Table 9.2 summarises the effect of pressure rating on material of construction. Except for lighter duties, rubber-impregnated fabric or laminated fabric construction is generally required for high pressure hydraulic systems. The question of a suitable type of seal is a matter of seal design and performance (and the manner in which it is used). The main requirements as far as the fluid is concerned (after ensuring that the pressure rating is adequate) are:-

(i) The fluid should provide adequate lubrication for dynamic seals.

(ii) Fluid and seal material must be compatible.

(iii) The seal material must have a maximum continuous service temperature at least equal to, and preferably considerably higher than, the maximum working temperature of the fluid.

(iv) The seal material must have suitable low temperature properties so that it does not become brittle or rigid when the system is standing idle at low ambient temperatures, or has to work at low temperatures.

TABLE 9.1
HIGH PRESSURE SEALS (RECIPROCATING SERVICES)

Type of Seal	Pressure psi				
	1 000	2000	3 000	5 000	10 000
PACKINGS: Metal Foil	X				
ASBESTOS (Plated)	X	X			
ASBESTOS (Proofed)	X	X	X		
ASBESTOS (Wire)	X	X			
COTTON (Bonded)	X	X			
O-RINGS	X				
O-RINGS (Backed up)	X	X	X		
U-RINGS (Rubber)	X				
U-RINGS (Fabric)	X	X	X	X	Possibly
V-RINGS (Rubber)	X	X	X	X	X
V-RINGS (Fabric)	X	X	X	X	X
C-RINGS	X	X	X	X	X
RUBBER/FABRIC COMPOSITE RINGS	X	X	X	X	Possibly
GLAND RINGS	X	X	X	X	X
CUSHIONED RINGS	X	X	X	Possibly	
ALL-METAL	X	X	X	X	X

TABLE 9.2
PRESSURE RATING FOR SEAL MATERIALS
(Pressure Rating in psi)

Type of Seal	Homogeneous Rubber	Semi-Reinforced	Rubber-Impregnated Fabric	Laminated Fabric
O-Rings, Quad Rings, etc.*	1 000—1 500	3 000	—	—
U-Rings	500—1 000	1 000	1 000—3 000	Up to 5 000
V-Rings	500—1 000	1 000—2 000	2 000—5 000	5 000—10 000
'Automatic' Sections	500—1 000	1 000—2 000	2 000—5 000	5 000
Composite Rings	—	Up to 3 000	Up to 10 000	—
Gland Rings	50— 150	250— 500	1 000—5 000	2 000—5 000

*Note pressure rating must also be related to type of service.

Item (i) is seldom an individual problem for a fluid which has sufficient lubricity to meet the requirements of the pump will almost certainly provide adequate seal lubrication. Adequate seal lubrication is also usually maintained at lower viscosities than would be tolerable, or desirable, in the pump, but in this case leakage past the seal may occur as a problem.

The only time when a seal is likely to receive inadequate lubrication is when static for a long time, and the seal dries out. This will appreciably increase the break-out force, and if the seal is very dry, may even cause damage to the seal. Provided the seal material is normally compatible with the fluid the most damaging condition for a pressure-energised seal is normally a very long idle period after installation in the system.

Item (ii) is a major problem, and requires careful selection of the seal material. Items (iii) and (iv) are related to service conditions, and may be regarded as imposing further limitations on material selection.

Compatibility

Compatibility can be defined, generally, as acceptable limits of swell and softening of an elastomeric material in contact with a fluid, with an absence of destructive attack. For most hydraulic applications it can be expressed specifically in terms of the swelling of an elastomer sample under continual immersion in the fluid at a specified temperature for a specified time. This empirical method has been standard for a number of years, acceptable swell levels being specified for individual industries, or particular requirements. More recently the method has been rationalised by the AHEM to yield a Seal Compatibility Index for mineral hydraulic oils which forecasts the swelling characteristics of the oil by tests on a standard rubber.

Here are thus two methods of assessing seal material compatibility

(i) The direct method where a specimen seal (or seal material) is tested for swell (and other signs of change or deterioration, if required) by immersion in a particular fluid – see Table 9.3.

(ii) The indirect method where the fluid is assessed for compatibility with a standard elastomer. This can be extended, and the performance of other elastomers estimated, by deriving a Seal Compatibility Index for the fluid, based on the test performance.

The former would normally be used for specific or special applications, and in particular where a special or non-standard fluid is to be employed in the system. The latter would be used for general industrial hydraulics, or specific systems, where conventional standard elastomers are used for seals, with conventional or standard oil fluids.

TABLE 9.3
DTD SPECIFICATION METHOD FOR DETERMINATION OF THE SWELLING OF SYNTHETIC RUBBERS

Three square test-pieces of the Standard Synthetic Rubber five grams in weight, shall be smoothly buffed and cleaned with a soft cotton cloth moistened with anhydrous ethanol or methanol. The volume of each test-piece shall be measured by determination of water displacement at room temperature, the weights in air and water being recorded to the nearest milligram. The test-pieces, after being dried by wiping with a soft cotton cloth, shall then be immersed in the material and in the High Swell and Low Swell Reference Oils for 168 hours, the temperature of the oils being maintained at $70^{\circ} \pm 1^{\circ}$ C. ($158^{\circ} \pm 2^{\circ}$ F.). At the end of this time the test-pieces shall be removed, the surface cleaned in anhydrous ethanol or methanol and wiped with a soft cotton cloth. Within five minutes of wiping the water displacement of each test-piece shall be again determined.
The increase in volume of each test-piece shall be calculated as a percentage from the initial and final water displacements.
The determination shall be carried out in duplicate.

Another method which has been widely employed in the case of mineral oil fluids is to use the aniline point as a measure of compatibility with elastomers. The higher the aniline point, in general, the better the compatibility is likely to be – i.e. the less the swelling and softening of rubbers. Use of the aniline point, however, is unreliable, even as a general guide, although the higher the aniline point of the oils used in the blend the better the compatibility should be with standard elastomers. Another parameter sometimes used – the viscosity gravity constant – is even more unreliable for predicting rubber swell.

The AHEM Seal Compatibility Index (SCI) is derived by immersing a standard square section test ring of medium low nitrile rubber for 24 hours in the oil held at a temperature of 100 degrees C. Before immersion the diameter of the ring is measured with a plug gauge. After the period of immersion the ring is allowed to cool and its diameter re-measured. The percentage change in diameter is then calculated, from which the Seal Compatibility Index can be derived by reference to tables – e.g. Table 9.4. Such a rating enables seal manufacturers to recommend the correct grade of rubber for use with any particular oil for which the SCI is specified, equipment manufacturers to specify the SCI for suitable oils; and oil manufacturers to produce blends with specific SCI ratings or ranges.

The overall subject of compatibility can now be considered in more general terms, dealing with fluids and materials separately.

Fluid Compatibility

Mineral oil fluids are compatible with a wide range of synthetic rubbers, but cause excessive swelling, deterioration and eventually

TABLE 9.4
CONVERSION OF LINEAR SWELL TO SEAL COMPATIBILITY INDEX

S_D %	D_1-S_1 inches	SCI	S_D %	D_2-D_1 inches	SCI	S_D %	D_2-D_1 inches	SCI
0.1	0.001	0.3	5.1	0.051	16.1	10.1	0.101	33.5
0.2	0.002	0.6	5.2	0.052	16.4	10.2	0.102	33.8
0.3	0.003	0.9	5.3	0.053	16.8	10.3	0.103	34.2
0.4	0.004	1.2	5.4	0.054	17.1	10.4	0.104	34.6
0.5	0.005	1.5	5.5	0.055	17.4	10.5	0.105	34.9
0.6	0.006	1.8	5.6	0.056	17.8	10.6	0.106	35.3
0.7	0.007	2.1	5.7	0.057	18.1	10.7	0.107	35.7
0.8	0.008	2.4	5.8	0.058	18.4	10.8	0.108	36.1
0.9	0.009	2.7	5.9	0.059	18.7	10.9	0.109	36.4
1.0	0.010	3.0	6.0	0.060	19.1	11.0	0.110	36.8
1.1	0.011	3.3	6.1	0.061	19.5	11.1	0.111	37.1
1.2	0.012	3.6	6.2	0.062	19.8	11.2	0.112	37.5
1.3	0.013	3.9	6.3	0.063	20.2	11.3	0.113	37.9
1.4	0.014	4.3	6.4	0.064	20.5	11.4	0.114	38.3
1.5	0.015	4.6	6.5	0.065	20.9	11.5	0.115	38.6
1.6	0.016	4.9	6.6	0.066	21.2	11.6	0.116	39.0
1.7	0.017	5.2	6.7	0.067	21.5	11.7	0.117	39.4
1.8	0.018	5.5	6.8	0.068	21.8	11.8	0.118	39.8
1.9	0.019	5.8	6.9	0.069	22.2	11.9	0.119	40.2
2.0	0.020	6.1	7.0	0.070	22.5	12.0	0.120	40.5
2.1	0.021	6.4	7.1	0.071	22.9	12.1	0.121	40.9
2.2	0.022	6.8	7.2	0.072	23.2	12.2	0.122	41.3
2.3	0.023	7.1	7.3	0.073	23.6	12.3	0.123	41.7
2.4	0.024	7.4	7.4	0.074	23.9	12.4	0.124	42.1
2.5	0.025	7.7	7.5	0.075	24.3	12.5	0.125	42.4
2.6	0.026	8.0	7.6	0.076	24.6	12.6	0.126	42.8
2.7	0.027	8.3	7.7	0.077	24.9	12.7	0.127	43.2
2.8	0.028	8.6	7.8	0.078	25.3	12.8	0.128	43.5
2.9	0.029	9.0	7.9	0.079	25.6	12.9	0.129	43.9
3.0	0.030	9.3	8.0	0.080	26.0	13.0	0.130	44.3
3.1	0.031	9.6	8.1	0.081	26.3	13.1	0.131	44.7
3.2	0.032	9.9	8.2	0.082	26.7	13.2	0.132	45.1
3.3	0.033	10.2	8.3	0.083	27.0	13.3	0.133	45.5
3.4	0.034	10.6	8.4	0.084	27.4	13.4	0.134	45.8
3.5	0.035	10.9	8.5	0.085	27.7	13.5	0.135	46.2
3.6	0.036	11.2	8.6	0.086	28.1	13.6	0.136	46.6
3.7	0.037	11.5	8.7	0.087	28.5	13.7	0.137	47.0
3.8	0.038	11.8	8.8	0.088	28.8	13.8	0.138	47.4
3.9	0.039	12.2	8.9	0.089	29.2	13.9	0.139	47.8
4.0	0.040	12.5	9.0	0.090	29.5	14.0	0.140	48.2
4.1	0.041	12.8	9.1	0.091	29.9	14.1	0.141	48.6
4.2	0.042	13.2	9.2	0.092	30.3	14.2	0.142	49.0
4.3	0.043	13.5	9.3	0.093	30.6	14.3	0.143	49.4
4.4	0.044	13.8	9.4	0.094	31.0	14.4	0.144	49.7
4.5	0.045	14.1	9.5	0.095	31.3	14.5	0.145	50.1
4.6	0.046	14.4	9.6	0.096	31.7	14.6	0.146	50.5
4.7	0.047	14.8	9.7	0.097	32.0	14.7	0.147	50.9
4.8	0.048	15.1	9.8	0.098	32.4	14.8	0.148	51.3
4.9	0.049	15.4	9.9	0.099	32.8	14.9	0.149	51.7
5.0	0.050	15.8	10.0	0.100	33.1	15.0	0.150	52.1

cracking of natural rubbers. Nitrile rubbers are the normal choice of elastomer for seals, unless high temperature working demands the use of more exotic synthetics. The maximum working temperature of nitrile rubbers is, however, higher than the maximum working temperature normally acceptable for oil fluids.

Fluid compatibility requirements may be specified as maximum permissible swell, e.g:-

Mineral oil fluids, in general, have little or no softening or stripping action on conventional paints and finishes, thus painting the inside of the reservoir is permissible.

Mineral oil fluids have no corrosive attack on metals and can thus be used in contact with virtually any metal. The presence of water in the oil will, however, tend to promote rusting of ferrous metal surfaces during idle periods, unless the oil is suitably inhibited (or the metal surfaces protected by painting). Water in oil can also cause corrosion of magnesium and magnesium alloy and so such metal should not be used for components in the system.

Castor-base fluids are compatible with natural rubber seals, and are normally used only with these. Castor-base fluids have a stripping action on normal paints, and thus special resistant paints must be used inside reservoirs, or the reservoir left unpainted. Magnesium should be avoided in contact with castor-base fluids as, in the presence of water, attack can be quite rapid. Otherwise castor-base fluids are essentially non-corrosive to metals. Compatibility with synthetic materials, including elastomers, may be limited, however.

Water-in-oil emulsions are generally similar to mineral base oil as regards compatibility with synthetic elastomers, and conventional seal materials are usually suitable. The presence of water does, however, increase the corrosion potential of the fluid – see also Chapter 8.

Water-glycol solutions are usually reasonably compatible with conventional seals and both nitrile and chloroprene types can be used. Water-glycol solutions are also compatible with natural rubbers, enabling such seals to be used for low temperature operation whilst retaining excellent flexibility.

Water-glycol solutions are not compatible with conventional paints and finishes, and also promote corrosive attack or staining of some metals – see Chapter 8.

Phosphate-ester fluids can be regarded as completely incompatible with conventional seals and thus require a complete change of seals, diaphragms and other elastomeric materials used in the system. Choice of elastomer is virtually restricted to butyl or ethylene propylene rubbers, or Viton. The latter, incidentally, can be used with both mineral oil and phosphate ester fluids.

Phosphate ester fluids have a stripping action on all types of paint finishes and limited attack on some metals – see also Chapter 8.

Corrosion characteristics of Fluids

The corrosiveness of fluids on metals can be assessed by immersing test strips of various metals in the fluid for a given period and assessing the effects. This can be a purely qualitative assessment – eg specified in terms of staining or visible attack; or quantitative, based on the change in weight and/or microscopic examination of the metal specimen. The quantitative method is used where a very high standard is set for the fluid requirements, e g for aircraft hydraulic fluids, to determine both the corrosiveness of the fluid and its resistance to oxidation. The following is the test method specified for fluids conforming to DTD 585 specifications.

(a) The apparatus shall consist of a large test-tube of heat resisting glass about 65 millimetres outside diameter and about 300 millimetres long, fitted by means of a ground glass connection to a water-cooled reflux condenser, and an air delivery tube of heat resisting glass about 6 to 8 millimetres in outside diameter and at least 50 millimetres longer than the total length of the test-tube and condenser when assembled One end of the air delivery tube shall be drawn out to a round aperture $^1/_{16} \pm ^1/_{64}$ inch internal diameter .

Test pieces about 1 inch square of copper (BS 899) low carbon steel (DTD 124A) aluminium alloy (BS 5L.3) magnesium alloy (DTD 118A) and cadmium plated steel (DTD 904) shall be cut from thin gauge sheet and drilled in each corner with $^1/_{16}$ inch holes. The cadmium plated steel test piece which shall have been plated after cutting and drilling shall be cleaned by rinsing in benzene, and the other four test pieces shall be cleaned and polished according to the method described in I.P.64.

The test pieces shall be weighed individually to the nearest milligram and tied together in the form of a pentagon by means of a cotton cord which has been previously boiled in distilled water and dried. The order of the metals in the pentagon shall be magnesium alloy, aluminium alloy, copper, cadmium plated steel and steel.

The metal pentagon shall be placed in the test-tube and the whole weighed to the nearest 0.1 gram. A volume of about 100 ml of the material shall be introduced, and the whole weighed again to the nearest 0.1 gram.

The test-tube shall be connected to the reflux condenser and placed in a thermostatically controlled liquid bath maintained at 121 degrees ± 1 degree C (250 degrees ± 2 degrees F). The air delivery tube shall be adjusted down the central tube of the reflux condenser so that its lower end with the $^1/_{16}$ inch aperture shall be within $^1/_8$ inch of the bottom of the test-tube and centrally within the metal pentagon. Clean, dry air shall be passed through the air delivery tube at the rate of 5 ± 0.5 litres per hour.

After 168 hours, the aeration and heating shall be stopped and the apparatus disconnected. The test-tube and contents shall be removed from the bath, cooled to room temperature, cleaned externally and weighed to the nearest 0.1 gram. Any loss by evaporation shall be calculated as a percentage of the weight of material originally taken, and in the event of evaporation loss being in excess of 5 per cent, the determination shall be disregarded, and a new determination made after the cause of excessive vapour leakage shall have been found and corrected.

Without removing the material from the test-tube,the metal pentagon shall be withdrawn and taken apart. The test pieces shall be rinsed first in benzene, then in acetone, dried and weighed individually to the nearest milligram. The change in weight of each test piece shall be calculated in milligrams per square centimetre of metal surface exposed to the material.

The test pieces shall be examined microscopically under a magnification of 15 diameters for visual evidence of pitting or etching.

The material remaining in the test-tube shall be examined visually for evidence of deposition of gummy or other insoluble matter, and determinations shall be made of its viscosity at 54.4 degrees C (130 degrees F) and of its neutralisation value.

(b) A strip of pure sheet copper, cleaned and polished according to Method I.P.64 shall be placed in a clean test-tube, and sufficient of the material added to immerse the strip completely. The tube shall be closed with a vented stopper and placed in a bath maintained at 100 degrees ± 1 degree C (212 degrees ± 2 degrees F) for 3 hours. The strip shall then be washed with sulphur-free acetone, and compared with a similar strip of freshly polished copper. Evidence of discolouration or pitting shall be determined by microeoscopic examination, using a magnification of 15 diameters.

Elastomer Compatibility (see also Tables 9.5 and 9.6)

Buna S (also known as GRS and SBR)

This synthetic rubber is a direct substitute for natural rubber, with somewhat inferior properties. It can be used as an alternative to natural rubber for seals, etc, used with castor-base fluids.

TABLE 9.5

COMPARATIVE PROPERTIES OF ELASTOMERS

Property	Neoprene	Natural rubber	Nitrile rubber	Butyl rubber	Ethylene Propylene rubbers	Polyeurethane rubbers	Viton
Tensile strength							
(a) Gum	3000+	3000+	1000	2500	1100–2000	4000–6000	
(b) Black loaded	3000+	3000+	2000	2500			
Hardness (Shore A or BS)	40–95	30–90	30–95	40–85	25–85	84–98	65–80
Tear resistance	Fair to good	Good	Fair	Fair to good		Outstanding	
Abrasion resistance	Good	Excellent	Excellent	Fair to Good		At least 5 times that of natural rubber	
Hydrocarbon resistance							
Aliphatics	Good	Poor	Excellent	Poor		Excellent	
Aromatics	Fair	Poor	Good	Poor to fair		Good	
Petrol and oil resistance	Good	Poor	Excellent	Poor		Excellent	
Acid and alkali resistance	Good	Fair to good	Good	Excellent		Poor	
Water resistance	Fair	Excellent	Fair to Good	Good	V. Good	Good up to 40°C (105°F)	
Gas permeability	Good	Fair	Fair	Excellent		Good to excellent	
Ageing: 1. Ozone	Excellent	Fair	Poor	Excellent	Outstanding	Outstanding	
2. Sunlight	Very good	Poor	Poor	Excellent	Outstanding	Good to excellent	
3. Oxidation	Good	Good	Fair to good	Good	Excellent	Excellent	
Service Temperature °C	75–100	75	75–100	75–100	50–120	75	200
°F	160–212	160	169–212	160–212	175–250	160	400
Max. Service Temperature	110°C	100°C	110°C	110°C	160°C	100°C	250°C
Flame Resistance	Good	Poor	Poor	Poor	Poor	Poor	Excellent
Dynamic Properties	Fair	Fair	V.Good	V.Good	V.Good	Fair	V.Good

TABLE 9.6
STANDARD MATERIAL CHOICE FOR HYDRAULIC SEALS

Fluid	Seal Material(s)
Water	Nitrile rubber
Water-in-Oil	Nitrile rubber
Water/Glycol	Nitrile rubber
Mineral Oil	Nitrile rubber or Polyurethane
Castor Base (Vegetable oil)	Natural rubber or Ethylene Propylene
Phosphate Ester	Ethylene Propylene or 'Viton'
Silicone Fluids	Nitrile rubber or 'Viton'

Butadiene (BR)

Also a direct substitute for natural rubber with somewhat better low temperature properties than Buna S. It can also be used with castor-base fluids, but is little used for hydraulic seals.

Butyl

Suitable for use with vegetable-base oils, but attacked by mineral oils and aromatic solvents. It was originally the main material used for seals with phosphate ester fluids, until replaced by ethylene propylene rubbers which have superior properties.

Chloroprene (also known as Neoprene)

Not widely used as a seal material since its resistance to mineral oils is only moderately good, and it also tends to harden and stiffen at low temperatures. It may offer certain advantages for use with oil-in-water emulsions.

Ethylene Propylene (EPM)

Generally excellent compatibility with phosphate ester fluids, and also silicone fluids, EPM is now usually the first choice seal material for use with the former fluid. It is not suitable for use with mineral oils and thus can be considered a specialised seal material. A further copolymer, EPDM may also be considered as it has enhanced performance characteristics for certain services.

Fluorocarbons

The range of elastomers in this family have enhanced mechanical properties, particularly as regards high temperature performance. Maximum continuous service temperature is of the order of 400 – 450 degrees F (200 – 250 degrees C). The main limitation is the extremely high cost of such materials. The best known is Viton, which is suitable for high temperature seals with both mineral oil and phosphate ester fluids, as well as silicone fluids up to 450 degrees F (250 degrees C).

Nitrile Rubbers

These are the standard materials for seals used with oil fluids. Oil resistance increases approximately in direct proportion to the nitrile content, but low temperature flexibility decreases in similar proportion. Thus a low nitrile rubber would be chosen for low service temperatures, at the expense of some resistance to oil at higher temperatures. Nitrile rubbers are also suitable for use with water and water-base fluids, water and oil emulsions, ester-glycol mixtures and silicone fluids. They may have limited compatibility with castor-base fluids and are completely incompatible with phosphate ester fluids and chlorinated hydrocarbons.

Polyacrylic Rubbers (ACM)

These elastomers have excellent resistance to oil fluids with a maximum service temperature higher than that of nitrile rubbers. They may be used as a cheaper alternative to Viton for oil fluid temperatures damaging to nitrile rubbers, but not exceeding about 350 degrees F (175 degrees C).

Polyurethane

This elastomer has excellent resistance to mineral oils and is also characterised by excellent mechanical properties, particularly resistance to abrasion. It would not normally be considered as an alternative to nitrile rubbers except for low temperature services (where its retention of flexibility is comparable to that of natural rubbers); or where maximum resistance to abrasion is important.

Silicone Rubbers

All silicone rubbers have an extremely wide working temperature range. They retain excellent flexibility down to temperatures as low as -100 degrees F (-75 degrees C) and have a maximum service temperature rating for continuous operation of the order of 450 degrees F (230 degrees C). The chief disadvantage of straightforward silicones is poor mechanical properties, although these can be enhanced by compounding. Cost is also another major disadvantage. Resistance to oil fluids is also somewhat limited, but again improved by compounding.

Fluorinated silicones offer enhanced performance, at some reduction in service temperature range. The oil resistance of fluorinated silicones can be made equal to that of nitrile rubbers.

Leathers

Leathers must also be included as suitable seal materials, since they have been used since the earliest days of water-hydraulics. Modern leathers have enhanced properties due to developments in processing and can be made compatible with a range of fluids. Wax-impregnated leathers are generally preferred for water services, this method of treatment yielding a stiff and rather hard leather. Poly-sulphide impregnation is used for high pressure seals and packings, this method of treatment improving the mechanical properties of the leather and its chemical resistance. For oil fluids a surface treatment with acrylic emulsions may be preferred since this toughens and adds flexibility to the surface of the material without penetrating the body of the leather. Polyurethane impregnation is also used to improve flexibility and wear resistance when used with water, water-base and oil fluids.

The two chief advantages of leather are its ability to maintain good sealing with minimum wear when rubbing over relatively rough surfaces and its ability to absorb and retain lubricant. The former means that a leather seal can be used on a surface which is too rough to yield proper sealing with an elastomeric seal – eg a worn or corr-oded surface. Also being a soft material its abrading effect on rubbing surfaces is negligible. Main disadvantages are the limited mould-ability of the material, which limits its use to straightforward seal sections, such as U's, V's and cups; and lack of rigidity and 'stretch'.

Further compatibility data are summarised in the Table 9.7.

TABLE 9.7
LEATHER SEALS

Fluid	Impregnant	Piston Seals	Rod Seals
Water	(i) Triglyceride or Stearate Stuffing	U & V rings Cups	U & V rings Cups
	(ii) Polyurethane	As above	As above
	(iii) Paraffin Wax	Smaller rings or cups	Smaller rings or cups
Water-in-Oil Oils	(i) Polyurethane	U & V rings cups	U & V rings Cups
	(ii) Polysulphide	U & V rings	U & V rings
Water-Glycol	–	U & V rings Cups	U & V rings Cups
Phosphate Ester	–	Not suitable	Not suitable
Silicone Fluids	–	Not suitable	Not suitable

TABLE 9.8
TYPES OF HYDRAULIC SEALS

Type	Groove	Application's)	Remarks
Rectangular 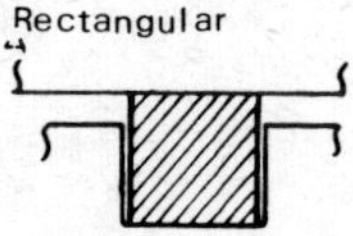	As for O-Rings	Alternative to O-Rings, particularly for slow speeds	Less backlash & more rigid than O-Rings. Higher friction.
Quad 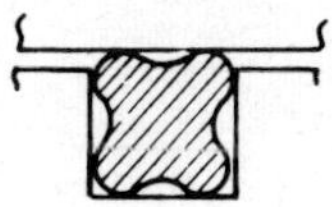	Generally closer fits than for O-Rings	Alternative to O-Rings, particularly for slow speeds.	More resistant to spiral failure at moderate pressures. Low backlash.
H 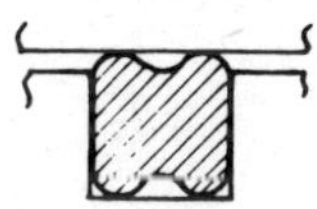	Generally closer fits than for O-Rings	Alternative to O-Rings, particularly for slow speeds.	More resistant to spiral failure at moderate pressures. Low backlash.
Delta 	Specific	–	High backlash. May also be used as back-up rings with larger O-Ring.
Nu-Lip 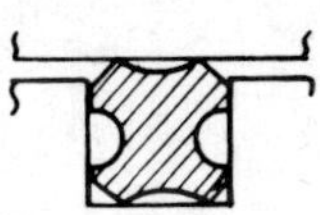	Similar to O-Rings	Alternative to O-Rings	Proprietary variant of the basic quad ring. Low backlash.
D 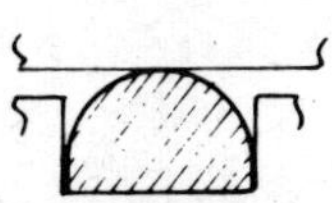	Specific	–	Good stability at low pressures. Resistant to spiral failure.
Trapezoid 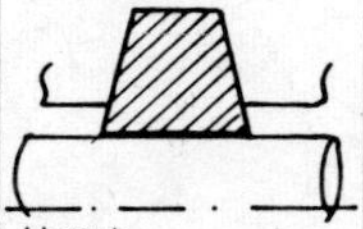	Close fit	Wiper rings	Also as lubricating rings employing porous material.
Heart 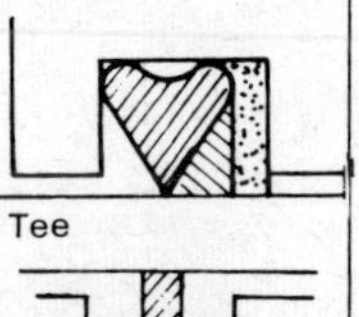	Standard O-Ring	Light Duty Rotary shaft seal only.	–
Tee 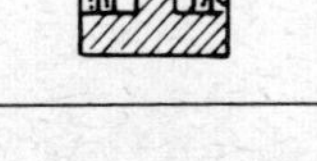	Specific	High pressures	Used with expandable back-up rings. High backlash.

TABLE 9.8 (continued)
TYPES OF HYDRAULIC SEALS

Name	Construction	Mechanics	Max. Working pressure psi (Standard tolerances)	Type(s)
Balsele 	Moulded rubber ring bonded to fabric base ring(s)	Preload by interference fit pressure energised	2 500	Single acting double acting internal/external
Grooved Balsele 	Lipped ring bonded to fabric base	Pressure energised & capable of accommodating eccentricity or side loads	2 500	Single acting internal/external
Dowty Composite 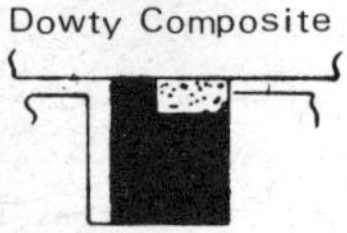	Hard PTFE insert in solid rubber ring	Preload by interference. Pressure energised.		Single acting (Internal & external) Double acting (Internal & external)
Hallprene 	Fabric U-ring & Rubber O-ring combination (also with rigid back-up rings)	Preload by squeeze pressure energised.	10 000	Single acting internal/external double acting internal/external
H.P. 	Rigid PTFE contact ring bonded to solid elastomeric ring	Contact pressure by squeeze. Pressure energised.	5 000–6 000	Double acting internal or external
Hypak 	Special rubber section bonded to nylon back-up ring	Squeeze and pressure energised		Single acting internal/external. Double acting internal /external
SCN 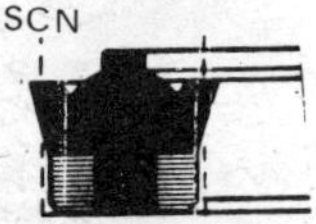	Special elastomeric section bonded to two concentric laminated rubber rings	Squeeze & pressure energised.		Single acting internal/external. Double acting internal /external
SCN Major 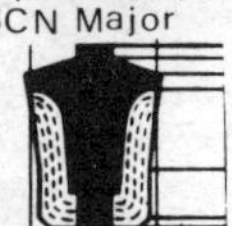	Special elastomeric section bonded to extended laminated fabric sections.	Squeeze & pressure energised.		Single acting internal/external
Solosele 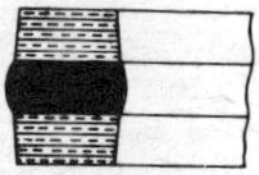	Modified O-ring section bonded to laminated fabric back-up rings	Squeeze & pressure energised.		Double acting only. Internal external.
TMS Hydraseal 	Special elastomeric section bonded to single laminated fabric ring on working face.	Squeeze & pressure energised.		Single acting Double acting internal or external.

TABLE 9.8 (continued)
TYPES OF HYDRAULIC SEALS

Type	Application	Characteristics	Remarks
Lip 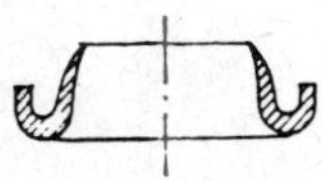	Flexible lip provides self-energising seal, preload pressure ensures static seal.	Friction fairly high at low pressures increasing with increasing pressure. Friction relieved on return stroke.	Basis of a number of proprietary seals. Also used in sets of similar seals, with automatic seals, etc., and headers.
Automatic 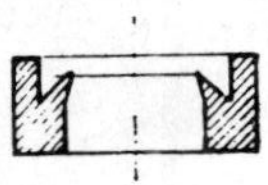	Self-energising seal, mainly for reciprocating motions.	Seal increases in efficiency with increasing pressure. Friction increases with increasing pressure.	Basis of a large variety of proprietary seal sections. Also used in pairs or multiple sets with or without headers.
U-Ring 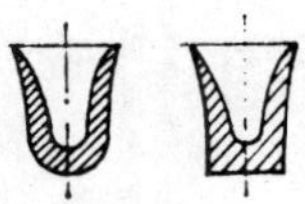	Piston Seals, etc, with or without follower ring. Generally suitable for pressures up to 1500 psi.	Good sealing at low pressures. High friction at high pressures. High wear at heel. Round or square base.	Many modifications of the shape produced as proprietary seals. Good general purpose seal where sliding motion is involved.
V. Ring or Chevron 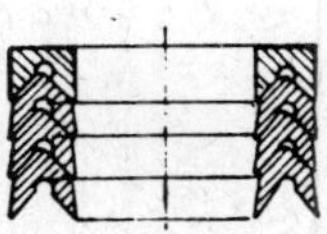	Reciprocating seals, gland seals, etc, where sufficient space is available. Very efficient and suitable for high pressures.	Light preload applied on assembly. Low friction at low pressures, increasing with increasing pressure.	Normally used in sets with support rings at each end. No. of Rings / Rubber psi / Rubber-Fabric psi / Leather psi 3 / Up to 500 / Up to 500 / Up to 500 4 / 500–1500 / 500–3000 / 500–2000 5 / 1500–3000 / 3000–5000 / 2000–6000 6 / 3000–5000 / –10000 / 6000–20000

TYPES OF HYDRAULIC SEALS TABLE 9.8 (continued)

Cup 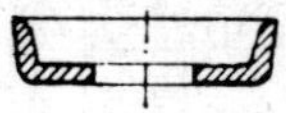	Almost exclusively used for piston seals. Available in a large range of standard sizes.	Excellent uni-directional sealing properties with reciprocating motions Low friction on return stroke.	Should be used with correct support plate and junk plate for optimum performance and maximum seal life.
Flange (Hat Washer) 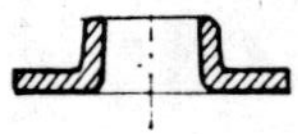	Widely used for internal sealing on shaft or piston at moderate pressures. Compact.	Preloading pressure may be increased with spring expander. More compact than U or V rings but less efficient as a seal.	May be used singly or in opposing pairs. Also used as a dirt or oil wiper ring. May be used with a gland ring, also expander spring.
Collar 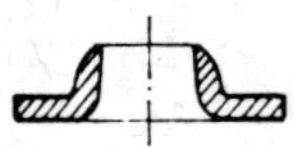	Similar to Flange rings.	Improved resistance to wear and scuffing by virtue of modified section	May also be used as wiper rings.
Seal Sets 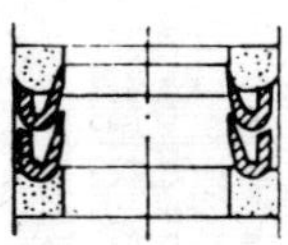	Gland and piston seals with or without wedging action.	Basis of seal sets is to provide sealing on both surfaces—pressure energised or self-energised.	Seal sets are widely employed with proprietary seals and sections.

TABLE 9.8 (continued)
TYPES OF HYDRAULIC SEALS

Name	Type	Application	Construction	Pressure Rating Normal psi	Max. psi	Remarks
Beldam 'Autolip'	Automatic	Ram packing	Cotton-Rubber Flax-Metal	—	6000	Suitable for temperatures up to 280°F.
Gaco Distributor Seal	Automatic	Internal or external piston or rod seal.	Homogeneous synthetic rubber.	—	—	Patent form of automatic seal.
F—VU F—VH	Chevron	Shallow glands	Moulded PTFE	350	1000	Modified V-section seal.
Lion Automatic	Automatic	High Duty Packing	Rubber impregnated Fabric	—	Up to 10000	Improved V-ring, used in sets.
Lion 'Twin-set'	U-Ring Set	Reciprocating	Laminated fabric or rubber header in hard composition.	—	10000	Used with spring loaded V-fluted header.
Rotopac (Polypac)	Automatic	Rotary	Tongued seal usable in sets	—	—	Particularly intended for centrifugal pump glands.
Seatrist Gland Ring	Lip	Reciprocating	Laminated fabric rubber impregnated.	—	10000	Used in sets with headers.
Seatrist TRU Ring	U-Ring	General purpose rod or piston seal.	Homogeneous moulded rubber.	—	—	Improved U-Ring design.
Seatrist WTG	V-Ring (Modified)	High duty packing	Rubber impregnated fabric	—	10000	Improved form of V-Ring, used in sets.

These Tables are reproduced from the Hydraulic Handbook - IVth Edition.

10 EVALUATION AND TEST METHODS

The tests or spot checks that the user can make on the standard of a hydraulic fluid are very limited, without access to an analytical laboratory. Even then, quantitative results so obtained may only be meaningful if a knowledge of the exact make-up of the fluid is known, e.g. the specific additives and their proportions. As far as the system user is concerned, therefore, fluid tests are limited to –

(i) Elementary qualitative tests

(ii) Simple tests specified by the fluid manufacturer – e.g. viscosity measurement to determine the water content of a water-base fluid by reference to data supplied by the manufacturer of that fluid.

Elementary qualitative tests may be made by withdrawing a sample of the fluid from the system and comparing the colour with that of a fresh fluid of the same grade –

Lighter colour, but still clear – would indicate that the fluid has possibly been diluted with a thinner oil. If so, this will probably have reduced the viscosity. This could be checked by a viscosity check.

Darker colour, but still clear – would indicate the onset of oxidation or the presence of fine contaminants. The degree of deterioration would be indicated by the degree of darkening, compared with the colour of a fresh sample. If the colour is removed by filtering, then contamination is present, indicating that the filters are inefficient. All the fluid should be removed from the system and filtered with a batch filter, and filter elements changed or replaced with more efficient types.

Darker colour, with cloudiness – more advanced deterioration of the fluid, or a higher degree of contamination. If the sample is rendered clear and light coloured again by filtering, then contamination may be the major failing and require treatment as above.

Smell can be a further check, especially in the case of a dark coloured oil sample. If rancid or 'bad' then advanced oxidation or degredation of the fluid can be suspected.

Cloudiness with no colour change – this can be an indication of water contamination in an oil fluid, but not necessarily so. Some oils develop cloudiness on standing (e.g. when the system is idle) which is not harmful and disperses when the fluid is agitated. Some fluids are naturally cloudy in appearance, e.g. oil-in-water emulsions, water-in-oil emulsions and some water-glycol mixtures. In the case of an oil fluid which is cloudy, this cloudiness should disappear on standing if caused by water contamination, with water separating out at the bottom of the oil. If not, then the demulsifying agent in the oil is exhausted.

Samples of fluid for checking should preferably be drawn after the system has been in operation and the fluid has achieved its normal working temperature. A sample drawn from the working or pressurised side of the system will then be most truly representational. To check specifically for water contamination in a system, this should be done by withdrawing samples both from the bottom and top of the reservoir after the system has stood idle for a period. Comparison, or further testing, of the two samples will then indicate the degree of water contamination and the effectiveness of the demulsifying agent in separating the water from the fluid.

It is important to appreciate that no realistic figure can be given for the life of any particular oil fluid since this will depend on the operating conditions. Thus the true life of the fluid in a particular system can only be determined by withdrawing samples at periodic intervals and submitting these to the fluid manufacturer for quantitative analysis. Simple quantitative tests are usually valueless, unless sufficient data is available to relate such results to the specific formulation of the fluid concerned. This is illustrated by the following.

The acidity of an oil is usually expressed in terms of a neutralisation number, of the number of milligrams of potassium hydroxide required to neutralise the acidity of one gram of the sample. In the case of an alkaline oil, the amount of acid required to neutralise 1 gram of the sample is determined, and then the number of grams of potassium hydroxide necessary to neutralise that amount of acid. The rate of increase in the neutralisation number is then a measure of the progress of oxidation of the oil, provided it has not become contaminated from an outside source. The neutralisation number of new, uncompounded oils is usually less than 0.05. However, it may be considerably different in the presence of additives (usually higher) and so the neutralisation number itself has little meaning without a knowledge of the formulation of the oil.

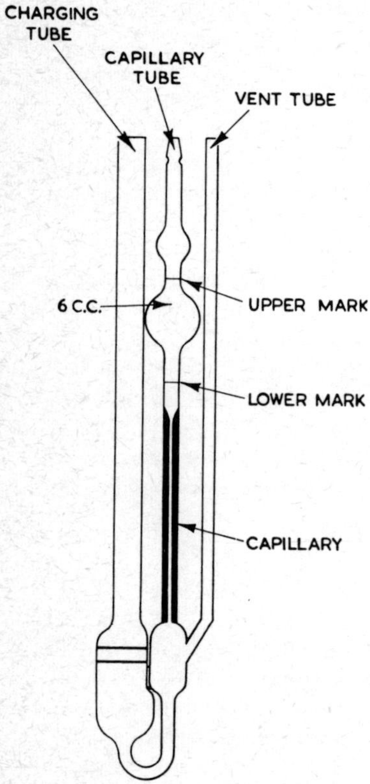

Fig. 10.1 Glass Viscometer

Viscosity Determination

Glass viscometers are generally used for the determination of kinematic viscosity of a fluid in absolute units. In these a sample of fluid is drawn up a capillary tube by vacuum (or forced up by pressure) and then allowed to fall under its own weight. The time is taken for the fluid level to pass between upper and lower calibrated marks on the capillary limb of the viscometer – Fig.10.1. The kinematic viscosity is then calculated from

$$\text{kinematic viscosity} = CT - \frac{B}{t}$$

where C = calibration constant for the instrument
t = time in seconds for the flow between marks
B = a constant determined by the capillary diameter and design of the instrument

For calibration purposes distilled water can be used as a 'standard' fluid, having a kinematic viscosity of 1.0068 centistokes at 68 degrees F (20 degrees C).

For practical measurement of viscosity in arbitrary units, one of the suspended level viscometers is generally used. Thus the Redwood instrument consists, basically, of a cup fitted with an agate jet through which the fluid can flow – Fig.10.2. The cup is surrounded by a water bath so that the fluid can be kept at a constant temperature, and a graduated flask is arranged below the jet to collect the fluid.

Tests are normally conducted at 70 degrees F, 140 degrees F and 200 degrees F. The procedure is to heat the fluid uniformly in a separate container to just above the test temperature and then pour it into the cup through a fine screen mesh. The fluid in the cup is then

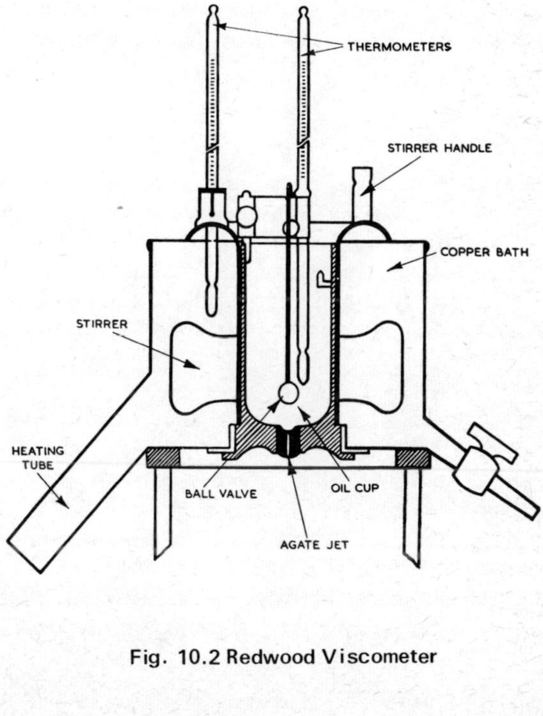

Fig. 10.2 Redwood Viscometer

stirred until it has cooled to the required test temperature, when a plug is removed from the jet allowing fluid to escape. The time is then measured for 500 ml of fluid to pass through the jet and collect in the graduated receiving flask. This time, in seconds, is quoted as the viscosity of the fluid at that temperature (e.g. seconds Redwood).

There are two forms of this instrument, the Redwood No.1 and Redwood No.2, differing only in detail and size of jet. The No.1 instrument is used for oils and other fluids having a viscosity not exceeding 2000 seconds at test temperature. The No.2 instrument has a flow rate 10 times faster than the No.1 and is normally only used for fluids having a viscosity of 1000 seconds or greater on the Redwood No.1 scale. Hydraulic fluid viscosities are invariably determined in Redwood No.1 seconds.

Corresponding standard instruments used by other countries are the Saybolt Universal (in America); and the Engler (in Continental Europe). The Engler system expresses viscosity in degrees (°E) instead of time, representing the ratio of time for 200 cc of fluid to flow from the viscometer to the time for 200 cc of water to flow out at 20 degrees C. Normally the time for 200 cc of water is 50 to 52 seconds, and so an 'equivalent seconds' value is given by multiplying Engler degrees by 50. Close approximate equivalents (for the same temperature) are given in the accompanying Table.

Viscosity Index

The slope of the viscosity-temperature curve for a particular fluid plotted on an ASTM chart is a definitive measure of the viscosity-temperature characteristics, often called the ASTM slope. However, since the scales are non-linear the ASTM slope is not a suitable way of assessing the rate of change of viscosity with temperature, and so this is done practically on a comparative basis, in terms of Viscosity Index (VI).

The Viscosity Index scale was originally established on the basis that a typical paraffinic oil showed a 'minimum' change in viscosity with temperature, and was thus given a VI rating of 100; whilst a typical naphthenic oil showed a 'maximum' change of viscosity with temperature and was given a VI rating of 0. The Viscosity Index of any other fluid is then determined on the basis of an 'equivalent mixture', i.e. the proportion of paraffin oil in a paraffinic-naphthenic mixture which would have the same viscosity-temperature characteristics as the fluid.

TABLE 10.1
VISCOSITY CONVERSIONS (AT SAME TEMPERATURE)

Centi-stokes	°E.	Redwood No. 1.	SUS	Centi-stokes	°E.	Redwood No. 1.	SUS
*1.0	1.00	28.5		20.0	2.9	86	97.5
*1.5	1.06	30		20.5	2.95	88	99.6
*2.0	1.12	31	32.6	21.0	3.0	90	101.7
*2.5	1.17	32	34.4	21.5	3.05	92	103.9
*3.0	1.22	33	36.0	22.0	3.1	93	106.0
*3.5	1.26	34.5	37.6	22.5	3.15	95	180.2
*4.0	1.30	35.5	39.1	23.0	3.2	97	110.3
*4.5	1.35	37	40.7	23.5	3.3	99	112.4
*5.0	1.40	38	42.3	24.0	3.35	101	114.6
*5.5	1.44	39.5	43.9	24.5	3.4	103	116.8
*6.0	1.48	41	45.5	25	3.45	105	118.9
*6.5	1.52	42	47.1	26	3.6	109	123.3
*7.0	1.56	43.5	48.7	27	3.7	113	127.7
*7.5	1.60	45	50.3	28	3.85	117	132.1
*8.0	1.65	46	52.0	29	3.95	121	136.5
*8.5	1.70	47.5	53.7	30	4.1	125	140.9
*9.0	1.75	49	55.4	31	4.2	129	145.3
*9.5	1.79	50.5	57.1	32	4.35	133	149.7
10.0	1.83	52	58.8	33	4.45	136	154.2
10.2	1.85	52.5	59.5	34	4.6	140	158.7
10.4	1.87	53	60.2	35	4.7	144	163.2
10.6	1.89	53.5	60.9	36	4.85	148	167.7
10.8	1.91	54.5	61.6	37	4.95	152	172.2
11.0	1.93	55	62.3	38	5.1	156	176.7
11.4	1.97	56	63.7	39	5.2	160	181.2
11.8	2.00	57.5	65.2	40	5.35	164	185.7
12.2	2.04	59	66.6	41	5.45	168	190.2
12.6	2.08	60	68.1	42	5.6	172	194.7
13.0	2.12	61	69.6	43	5.75	177	199.2
13.5	2.17	63	71.5	44	5.85	181	203.8
14.0	2.22	64.5	73.4	45	6.0	185	208.4
14.5	2.27	66	75.3	46	6.1	189	213.0
15.0	2.32	68	77.2	47	6.25	193	217.6
15.5	2.38	70	79.2	48	6.45	197	222.2
16.0	2.43	71.5	81.1	49	6.5	201	226.8
16.5	2.5	73	83.1	50	6.65	205	231.4
17.0	2.55	75	85.1	52	6.9	213	240.6
17.5	2.6	77	87.1	54	7.1	221	249.9
18.0	2.65	78.5	89.2	56	7.4	229	259.0
18.5	2.7	80	91.2	58	7.65	237	268.2
19.0	2.75	82	93.3	60	7.9	245	277.4
19.5	2.8	84	95.4	70			323.4

*NOTE: in this range the table gives reasonably accurate conversions from centistokes to arbitrary scale units, or for conversion between arbitrary scales; but not for conversion of arbitrary scale units into centistokes.

For viscosity values above the range covered by the table, conversion is best obtained by factoring, viz

°E = o.132 centistokes = 0.0326 Redwood No. 1 = 0.0285 S.U.S.

Redwood No. 1 = 4.05 centistokes = 30.7°E = 0.887 S.U.S.

S.U.S. = 4.62 centistokes = 35.11 °E = 1.14 Redwood No. 1.

centistokes = 7.58 °E = 0.247 Redwood No. 1 = 0.216 S.U.S.

More specifically:-

$$\text{Viscosity Index} = \frac{L - U}{L - H} \times 100$$

where L = viscosity at 100 degrees F of an oil of 0 VI (naphthenic) having the same viscosity at 210 degrees F as the fluid sample

U = viscosity of oil sample at 100 degrees F

H = viscosity at 100 degrees F of an oil of VI 100 (paraffinic) having the same viscosity at 210 degrees as the fluid sample

This formula enables the Viscosity Index of a fluid to be determined if its viscosity is known at two temperatures – 100 degrees F and 210 degrees F. Values of L and U are read from standard tables, consistent with the fluid viscosity at 210 degrees F.

This method is valid for determining VI values up to 100. Certain fluids (e.g. water), and oils with viscosity index improvers, have VI values considerably in excess of 100. Extra-polation of the standard method to cover VI values above 100 yields inconsistent results and so a modern extension of the system has been derived to calculate VI values over 100. These are correctly designated VI_E. Thus a VI_E value quoted for high viscosity indices can be taken as more truly indicative of the viscosity-temperature characteristics of a very high viscosity index fluid than a VI value.

The deviation of VI_E valves is given in ASTM method D.2270-64, and also the Substitute of Petroleum publication I.P.226/68. See also 'Viscosity Index Extension'.

Pour Point Measurement

The pour point of a fluid is determined empirically, using standard apparatus, as 5 degrees F (-15 degrees C) above that temperature at which the fluid ceases to flow – see Fig.10.3. It represents the highest practical viscosity value where the substance is still fluid, or in practical terms the lowest temperature at which the substance can still be considered to be a fluid.

Standard tests approach the pour point by progressive cooling in 5 degree F stages to where the fluid sample shows no signs of flow when the tube is held in a horizontal position. The pour point is then specified as 5 degrees F above this condition. Thus pour points are specified in multiples of 5 degrees F and are not strictly related to service or storage conditions where actual pour point may be significant.

Bulk Modulus

A diagrammatic representation of a bulk modulus tester developed by NEL is shown in Fig. 10.4. Commercial versions of this unit are available and enable measurements of fluid compressibility to be made more easily, and more rapidly, than with most other testers of this type.

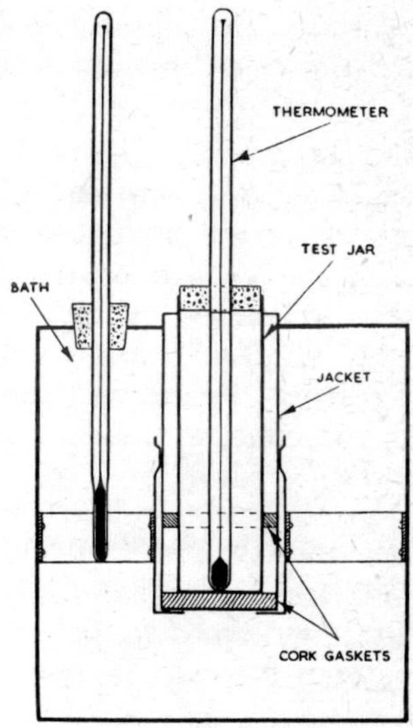

Fig. 10.3 Pour Point Apparatus

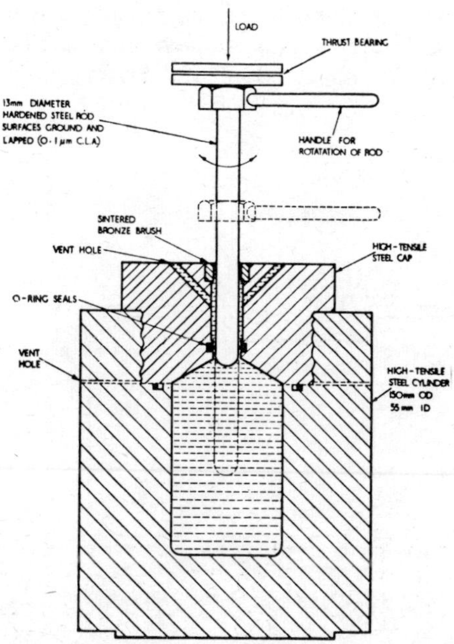

Fig. 10.4

The tester comprises, basically, a pressure vessel completely filled with the fluid under test and a steel rod which is forced into the vessel through an O-ring seal. By having a very smooth surface finish on the rod, and rotating it during the test, friction is reduced to an absolute minimum and leakage at the seal eliminated (the seal being changed after each test). The load on the plunger is thus directly proportional to the pressure in the vessel, and the distance that the plunger moves is proportional to the reduction in fluid volume. The actual reduction in fluid volume is determined by deducting an allowance for the expansion of the pressure vessel under the test pressure.

An autographic testing machine is used to force the rod into the pressure vessel, the load/displacement curve for this machine being converted into a pressure/volume curve for the fluid under test. Isothermal curves are obtained by applying the load very slowly, and isentropic curves by applying the load very rapidly. Corresponding times are approximately three hours and ten seconds, respectively. If isothermal tests are being made, or isentropic tests at other than room temperature, the pressure vessel is surrounded by a constant-temperature bath.

Shear Stability Measurement

One standard apparatus for the determination of the shear stability of a fluid is shown in Fig.10.5. Fluid is pumped and continuously circulated around the test circuit for a period of 168 hours, during

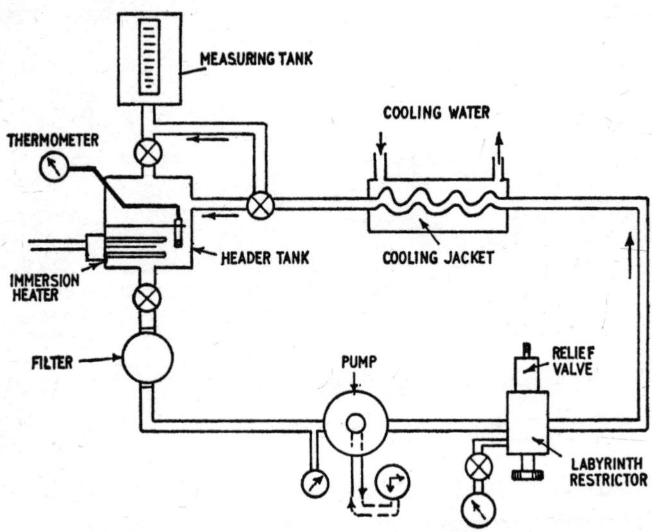

Fig. 10.5 Apparatus for shear stability test

which time the temperature in the header tank is maintained at 100 degrees F (38 degrees C), plus or minus 5 degrees F ($\pm$ 2 degrees C). At the same time the pump inlet suction is maintained at 3.0 inches Hg $\pm$0.5 in.Hg; the pump speed at 2500 rpm $\pm$ 50 rpm; and the relief valve pressure 800 psi $\pm$ 25 psi (56 kg/cm^2$\pm$1.75 kg/cm^2).

At the end of the test period the viscosity of the fluid is measured and compared with the original viscosity to determine the viscosity loss through shear breakdown, this being done at suitable reference temperatures. The fluid may also be examined for deterioration, separation, etc. Characteristics after the shear test may be specifically stated, or given as comparative to the performance achieved with a standard reference fluid.

Various other tests may be applied to determine the shear stability of a fluid,this being a subject investigated in detail by the A3 Working Party of AHEM in an attempt to introduce standardised laboratory tests. However, specific industries continue to specify their own standard requirements (e.g. the aircraft industry); and individual fluid manufacturers their own techniques.

The use of a diesel injector as a shearing device (i.e. restrictor) is probably the most satisfactory method of reproducing the shearing action which occurs in industrial hydraulic circuits and simplified rigs can be based on such restrictors – see Figs. 10.6 and 10.7. The test circuit need comprise no more than an electrically driven pump forcing fluid around a simple circuit and through a specified diesel ejector, with provision in the circuit for filtering and de-aeration. For short term tests fluid viscosity can be taken by sampling after 1, 10, 100 and 250 cycles through the injector.

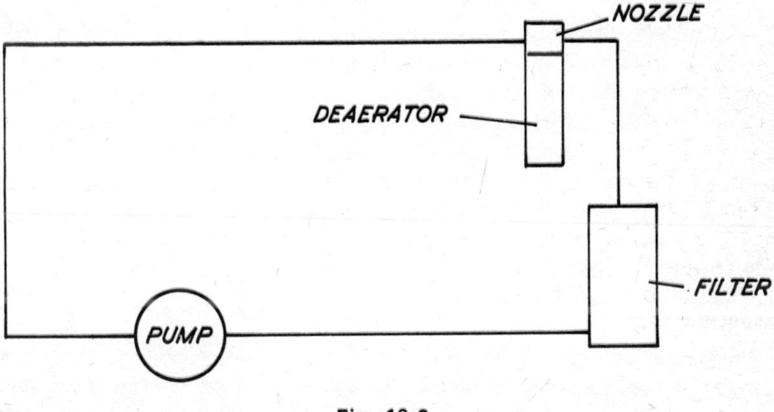

Fig. 10.6

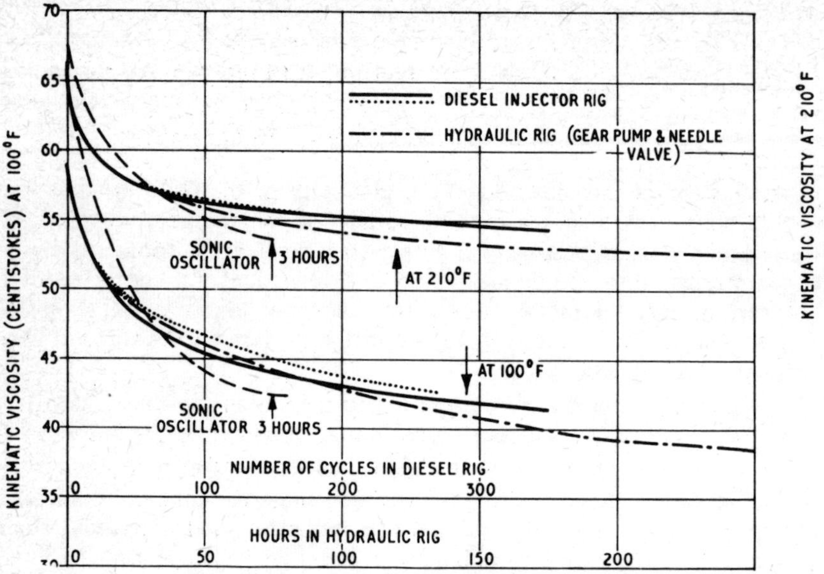

Fig. 10.7 Comparison of different shearing methods

Low Temperature Stability

This may be assessed in terms of the turbidity produced when maintained at a specified low temperature for a given period. In English practice the standard low temperature is usually -65 degrees F and the soak time a minimum of 72 hours.

A standard turbidity fluid is prepared by adding 125 ml of distilled water and 25 ml of 0.5 normal sulphuric acid to 25 ml of a 0.00322 molar solution of barium chloride. This mixture is well shaken, rendered alkaline by the addition of 25 ml of normal sodium hydroxide and made up to 250 ml with distilled water. Soluble red dye is then added to match the colour of the fluid sample. For satisfactory low temperature stability, the test sample after exposure should have no greater turbicity than that of the prepared standard solution so mixed and compared within thirty minutes of having been mixed.

Oxidation Tests

Numerous methods have been devised of measuring the oxidation or induction period of oils, although all suffer from certain limitations (some to the extent that they are of doubtful value). Agreement, too, is often poor between different methods of testing the same oil. Thus two of the most satisfactory methods only will be mentioned.

The *Turbine Oil Stability Test* is a long term test where oil is maintained at 95 degrees C in contact with copper and iron catalyst coils and subject to oxygen gas flow at the rate of 3 litres per hour. The neutralisation number is taken as a measure of the degree of oxidation and the useful life of the oil is taken as the time to reach a neutralisation number of 2.0 – see Fig.10.8. Test time may run to several days, or even weeks.

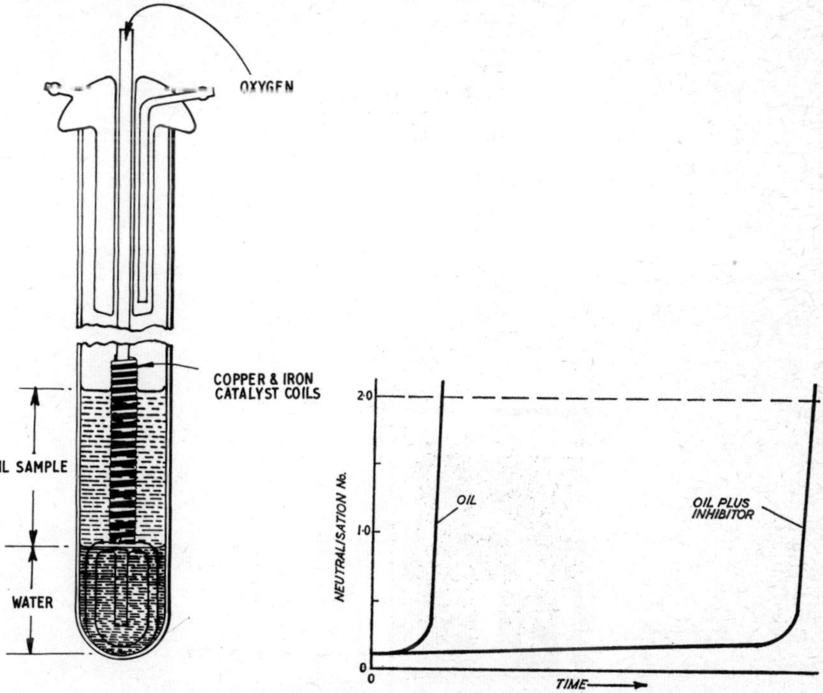

Fig. 10.8 **TURBINE OIL OXIDATION STABILITY TEST**

To assess resistance of turbine oils to Oxidation

Condenser	**'Mushroom'**
Oil	**300 ml**
Water	**60 ml**
Catalyst	**3 metres of iron and of copper wire as spiral round the oxygen inlet tube**
Temperature	**Oil at 95 deg C**
Oxidising gas	**Oxygen**
Rate	**3 litres/hr**
Life of Oil	**Time required to reach neutralisation value = 2**

The *Shell Rotary Bomb Oxidation Test* is a short term test. The 60 ml oil sample in this case is floated on 20 ml of water and in contact

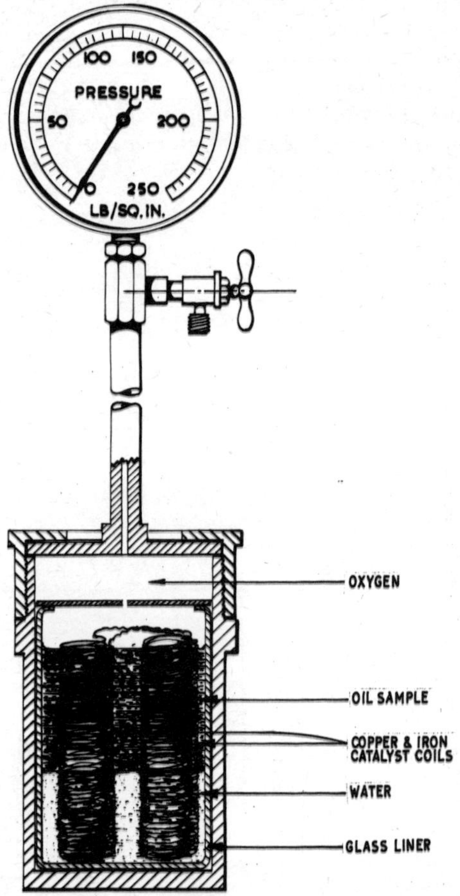

Fig. 10.9 TURBINE OIL ROTARY BOMB OXIDATION TEST

To assess resistance of Turbine Oils to oxidation (direct oxygen – absorbtion)

Oil	**60 ml**
Water	**20 ml**
Catalyst	**4 double iron/copper helices each containing 2½ feet of each metal**
Oxygen	**90 lb/sq in at 25 deg C (approx. 170 lb at 150 deg C)**
Oil Bath	**150 deg C**
Rotation	**100 rpm**

Life taken as Total Time to 'break' pressure minus half 'heating-up' time

with copper and iron catalyst coils in a pressure vessel. The coils are helix wound and each contain 30 inches of each metal – Fig.10.9. Pressurised oxygen is fed to the chamber, which is also heated up to 150 degrees C in an oil bath. Oxygen pressure is 90 psi (6.3 kg/cm²) at 25 degrees C, rising to approximately 170 psi (12 kg/cm²) at 150 degrees C. Oil life is taken as the total time to 'break' pressure minus half 'heating up' time. Such a test can normally be completed within a few hours, yielding curves similar to those shown in Fig.10.10.

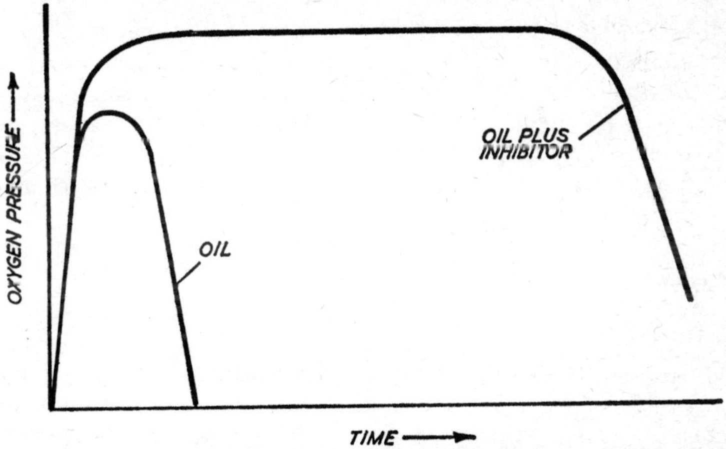

Fig. 10.10 Shell Bomb Oxidation pressure time curves. Life taken as total time to 'break' pressure minus half 'heating up' time This is a direct oxygen absorption test.

Other test methods include the AM oxidation test, primarily employed for crankcase oils and the Synder Life, Sligh Oxidation and Staeger Life tests used in America. Oxidation tests are of primary importance to oil fluid manufacturers, most of whom have their preferred, or specially developed methods. They are largely of academic interest only to the fluid user since the working life of a fluid in a system is largely governed by the operating conditions and working temperature of the fluid in the system, and thus can be expected to differ from laboratory test life.

'Lubricity'

'Lubricity' is a word coined to describe the lubricating properties of a fluid. Although not generally accepted – and held by some authorities to be misleading because it is unspecific – it is, nevertheless, a convenient term to use.

Since the pump in a hydraulic circuit normally sets the lubricity requirements of the fluid, the best practical measure of lubricity is the wear rate on the pump, which can then be expressed specifically, and comparatively for different fluids. This, however, is a long term test technique. Short term tests are needed for laboratory tests or standard tests on fluids covering the requirements of hydraulic pumps and motors. Numerous such tests have been developed, but co-relation between them is often poor. One basic limitation of many such simplified tests is inability to distinguish between the anti-wear properties of mineral oils with and without anti-wear additives.

Vickers Vane Pump Test

To classify oils for use in conditions of maximum severity the Vickers Corporation of Detroit have evolved a test in which a Vickers vane type pump, type 105C, is run at a speed of 1200 r.p.m., an oil pressure of 2000 p.s.i. with a flow rate of 5 gallons per minute, the oil temperature being maintained at 65° C., thus stimulating conditions of severe overload.

Measurements are made of wear in terms of loss of weight on the inner and outer port plates, cam ring and vanes, against bonus run - see 'Wear Tests'.

Aniline Point

The aniline point of an oil is determined using the apparatus shown in Fig. 10.11. Equal volumes of the oil and aniline are heated (or cooled) whilst stirring in a test tube and that temperature at which the mixture is just miscible is taken as the aniline point.

Aniline will be miscible according to the proportions of constituents with high solvent power, the aromatics having higher solvent power than the paraffins or naphthenes. Thus a low analine point indicates

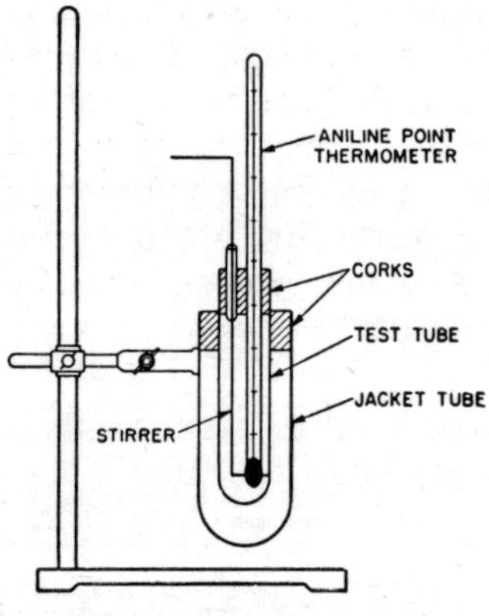

Fig. 10. 11 Detail of apparatus employed for the determination of aniline point

a high aromatic content, and vice versa; whilst the aniline point itself is an indication of the likely solvent action of the oil. More specifically the lower the aniline point the more likely the oil is to attack and cause swelling of rubbers.

Flash and Fire Points

'Flash' and 'fire' points indicate the temperature at which an oil begins to volatilise. The flash point, basically, is the temperature at which the oil will give off sufficient vapour to ignite momentarily. The fire point is taken as the temperature at which the vapour will ignite and continue to burn once combustion has been initiated by an external flame or spark.

In more specific terms the flash point is a measure of the inflammability of a liquid and, apart from assessment of fire hazard, can also be taken as a rough guide to oil blend, tendency to vaporise under high service temperatures, etc.

Various methods of test are used, the flash point obviously depending to a certain extent on the conditions of test. Open flash points correspond roughly to service conditions where oil may come into contact with a heated surface (e.g. dripping cuts on an exhaust manifold) and differ materially from closed flash points (roughly equivalent to the heating of oil in a closed system or storage container). Quoted without context, flash points have only comparative value. The Pensky-Martens closed cup test is generally employed for oils having a flash point higher than 120 deg. F (90 deg C) - see Fig.10.12. Flash point

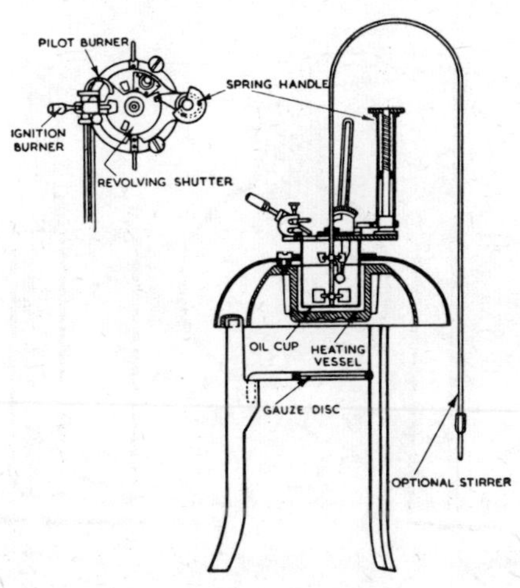

Fig. 10.12 Pensky - Martens flash point apparatus.

is determined to the nearest 20 deg F up to 220 deg F; and to the nearest 5 deg F for higher temperatures.

Foam Tests

Comparatively simple tests may be employed to assess the resistance of hydraulic fluids to foaming and their readiness to release air. These are generally based on measurement of the air retained after a test period – see Fig. 10.13.

Demulsification Tests

A standard test for the measurement of demulsibility or the ability of an oil to separate readily from water is ASTM 1401 Test; or, alternatively the Steam Emulsion Number Test IP19. Apparatus for measurement of the Steam Emulsion Number is shown in Fig.10.14. The test determines the time, in seconds, required for a 2 m litre volume of oil to separate from an emulsion formed with an equal volume of condensate, this time being quoted as the Steam Emulsion (SE) number of the oil.

Spontaneous Ignition Temperature

The 'spontaneous ignition temperature' (S.I.T.) is a higher temperature still at which a mixture of vapour and air will ignite without the application of an external spark or flame. It is totally unrelated to the

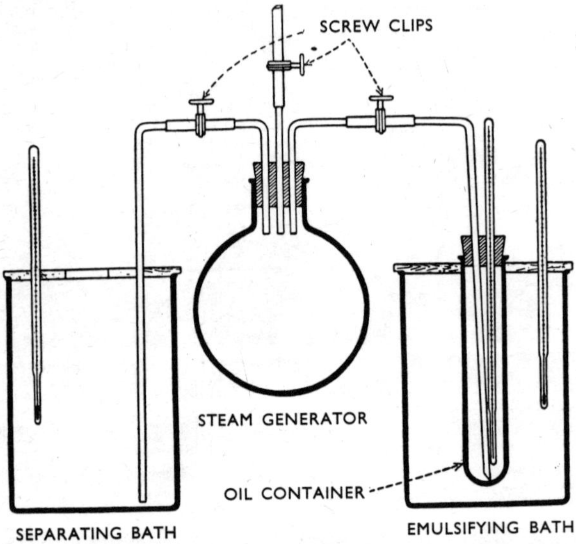

Fig. 10.14 Apparatus for determination of Steam Emulsion Number

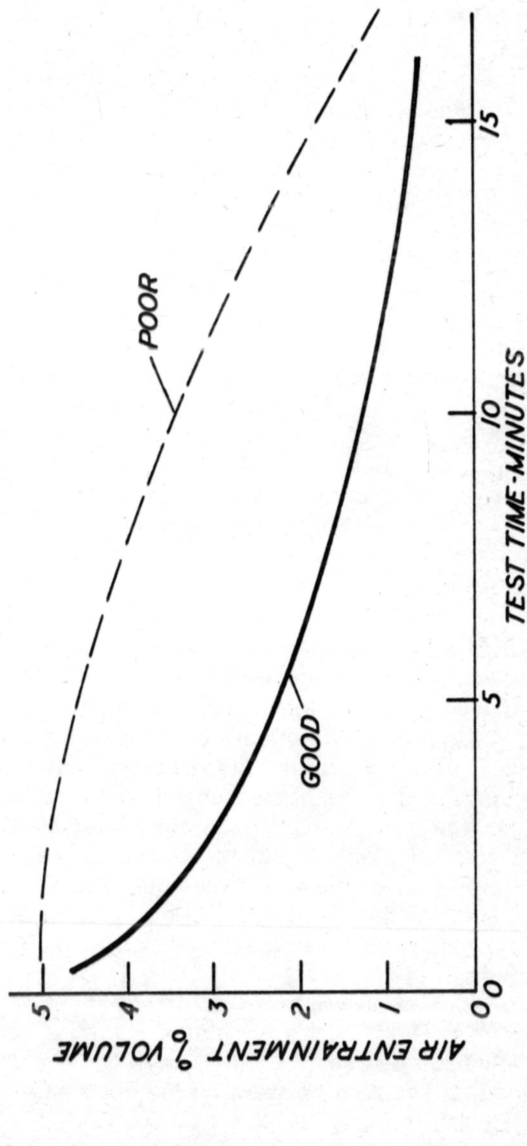
POOR
GOOD
AIR ENTRAINMENT % VOLUME
0
1
2
3
4
5
TEST TIME-MINUTES
0
5
10
15

Fig. 10.13

flash point and is also dependent on the vapour and air mixture being within the requisite explosive limits. It is of obvious significance in the case of oils in systems subjected to adiabatic compression where a 'diesel' explosion could be initiated when the S.I.T. of the fluid corresponds with the peak temperature realised.

Flammability Tests

No overall method exists for assessing fire-resistance, which can be defined, basically, as the ability to resist ignition by a heated surface or hot flame, plus an inability to support combustion or propagate a flame. Various types of flammability tests have, however, been introduced, both in military specifications for fire-resistant aircraft hydraulic fluids and by individual manufacturers as a standard for the development of their own products for industrial application. In the main these embrace flash point tests, dropping fluid onto a heated surface or a molten metal, spraying fluid onto a heated surface, investigating the ignition properties of a high pressure spray with a suitable flame(e.g. oxy acetylene)and the 'pipe cleaner' test. Performance may be evaluated against a known standard(e.g. a reference fluid which may not necessarily be a practical hydraulic fluid), or against specific requirements of non-ignition. Some, which are distinctly laboratory tests, may bear little or no relationship to ignition conditions likely to be met in practice and thus are mainly of academic interest. Others aim at duplicating practical conditions that could arise in the event of system failure giving rise to a fluid leak or fluid spray.

Flash point tests are of a class which do not significantly designate the potential flammability of a fluid since standard flash point tests are conducted under controlled conditions peculiar to the test method (and leading to a higher concentration of vapour than would normally be present under practical conditions). Many of the synthetic fluids, in fact, have (laboratory) flash points similar to mineral oils, although in practice they may be essentially non-flammable. Water-base fluids, too, have no conventional flash point but water-in-oil emulsions may exhibit apparent flash points on test due to separation of the oil content or boiling off of the water content. In this case, of course, it is the flash point of the oil content which is being determined, which bears no relationship to practical conditions which might be experienced with the emulsion.

The manifold test attempts to duplicate practical conditions by dripping the test fluid onto a simulated manifold section heated to approximately 1 300 degrees F (700 degrees C). A fluid is considered to pass this test if it does not burn on the hot manifold, or upon leaving the heated manifold.

The high pressure spray test aims at duplicating conditions equivalent to local damage to the hydraulic system resulting in a fine spray of fluid being forced out under pressure and reaching a naked flame. The spray is produced by forcing the fluid through a straight edged orifice of .0145 in at a pressure of 1 000 psi and then attempting to ignite the spray with an oxy-acetylene flame. A variation of this test employs a fuel-oil nozzle in place of the orifice, the nozzle being designed to promote combustion of fuels at 100 psi, or less. It is possible in this latter test that synthetic fluids and water-in-oil emulsions may be so effectively atomised at 1 000 psi pressure that they can be ignited by the torch flame.

A test to simulate industrial accident conditions is to pour the fluid onto the surface of molten metal – eg molten aluminium to simulate the conditions which could arise with aluminium die-casting machines. Behaviour in this case will be dependent on the duration of the test and the type of fluid used. Water-in-oil emulsions will show delayed ignition due to the 'snuffing' action of the steam formed, but in time the oil content may be increased to the point where it ignites readily and burns with a strong, even vigorous flame. Water-base fluids will, in general, exhibit inert properties, but again when the water content has been driven off in the form of steam the remaining glycol content may burn. Phosphate esters will not normally tend to burn at all although if sufficient time elapses for them to dissociate chemically the vapours may burn. In actual fact phosphate ester fluids themselves are essentially non-flammable but under extended heating some decomposition can take place and the resulting gaseous products of decomposition can burn.

The 'pipe cleaner' test is a simple, direct method of establishing an arbitrary standard of flammability, although not directly related to practical conditions. In this, an ordinary pipe cleaner is soaked in the fluid and then passed to and from through a flame until it catches alight. The number of passes required to produce ignition are taken as the rating of the fluid for 'flammability'. Typical results obtained by the 'pipe cleaner' tests are summarised in Table 10.2.

TABLE 10.2
"PIPE CLEANER" FLAMMABILITY TEST
(Typical test results)

Fluid	Number of Passes to ignite
Mineral oil	2 to 3
Phosphate ester	80
Phosphate ester/Chlorinated aromatic	80
Water-in oil emulsion	50
Water/Glycol	66

Specification Requirements

For industrial hydraulic systems the fluid specification is normally established by the fluid manufacturer, covering the number, type and proportion of additives thought necessary; and with various blends to cover a range of different viscosities. The user then simple selects a fluid of suitable viscosity, consistent with the pump manufacturer's recommendations, or other specific requirements of the system.

In the case of ultra high duty systems – e.g. particularly aircraft systems – fluids must be produced to a specific specification and requirements, based primarily on user experience. It does not necessarily imply that these are superior fluids, merely that the requirements for the fluid are more explicitly expressed. Many standard industrial grade hydraulic oils, in fact, may meet the majority of such requirements without conforming to the complete specification.

As an example of a specific material specification, the following are requirements for DTD 585 mineral blade hydraulic fluid for aircraft.

1. The material shall consist of refined petroleum oil with approved additives to ensure compliance with the requirements for viscosity and resistance to oxidation. The material shall contain tricresyl phosphate as an anti-wear agent. Pour point depressant additive shall not be used. The material shall be dyed red for identification purposes.

2. *Additives* (a) The viscosity-improver additive in the material shall not exceed 10 per cent by weight of active ingredient.

(b) The quantity of oxidation inhibitors in the material shall not exceed 2 per cent by weight.

(c) The quantity of tricresyl phosphate in the material shall be 0.5 ± 0.1 per cent by weight. The tricresyl phosphate shall comply with the requirements of D.T.D. 850.

3. *Freedom from Impurities*. The material shall be clear, free from dirt and other visible impurities.

4. *Viscosity*. The viscosity of the material shall not be less than 10 centistokes at 54.4 degrees C (130 degrees F) or more than 500 centistokes at minus 40 degrees C (minus 40 degrees F).

5. *Pour Point*. The pour point of the material shall be not above minus 59.4 degrees C (minus 75 degrees F).

6. *Flash Point*. The open flash point of the material shall be not below 93.3 degrees C (200 degrees F).

7. *Neutralisation Value*. The neutralisation value of the material shall be not more than 0.2.

8. *Colour*. The material shall contain oil soluble red dye, colour index 258, in concentration, not more than 1 part in 10,000 parts by weight of the material.

9. *Stability at Low Temperature.* The material after being maintained at a temperature of minus 54 degrees C (minus 65 degrees F) for 72 hours shall not solidify, gel, crystallise, or show signs of precipitation, and any turbidity present shall be not greater than that shown by the turbidity standard.

10. *Stability under Shear.* The stability of the material under shear shall be such that when determined by standard test technique the percentage change in viscosity of the material at 54.4 degrees C (130 degrees F) and at minus 40 degrees C (minus 40 degrees F) shall be not greater than that of the Shear Stability Reference Oil. The material shall not show visible evidence of darkening, separation of gummy or other solid substances, or sludge formation. The neutralisation value shall not have increased by more than 0.2.

11. *Swelling of Synthetic Rubber.* The swelling effect of the material on the Standard Synthetic Rubber when determined by standard test method shall be within the limits established by the Low Swell and the High Swell Reference Oils. There shall be no disintegration or deterioration of any of the rubber samples on the oil.

12. *Resistance to Evaporation.* The resistance of the material to evaporation shall be such that when determined by the standard method the tackiness of the film shall not be greater than that of a film of the Tackiness Standard.

Additional requirements are also set for corrosiveness and resistance to oxidation (see Chapter 9); and type approval is subject to inspection.

Water Content of Water-Glycol Solutions

Where specifically required, the water content of water-glycol solutions can be determined accurately by chemical analysis, using Fischer Reagent. This reagent consists of a methanolic solution of iodine and sulphur dioxide to which pyridine has been added. In the presence of water, iodine and sulphur dioxide react to form hydriodic acid and suplhur trioxide which, in turn, are neutralized by the pyridine. When a hydrolube sample is titrated with the reagent, the colour fades instantaneously until all the water is consumed. The persistence of the brown iodine colour indicates the end point of the reaction. Calculation then gives the per cent by weight of water in the sample.

The reagent is prepared by dissolving the following chemicals in the same order and approximate weight proportions given below. All ingredients should be substantially dry.

Methanol . . . . 5 litres
Iodine . . . . 254 grams

Pyridine . . . . 790 grams
Sulphur Dioxide . . . . 192 grams

Sulphur dioxide is weighed by passing the gas into a weighed receptacle containing the solution of methanol, iodine, and pyridine. The container is placed on a balance under a hood and the sulphur dioxide gas is passed into the solution until the desired weight has been added. The reagent should be allowed to stand for at least a day before standardization and should be standardized whenever used.

Since the reagent rapidly loses strength when exposed to air, it is best preserved and used in a 'pyrex' automatic, inter-joint burette which prevents direct contact of the liquid with moist air. All vents open to the air must have protective drying tubes. The accuracy and reproducibility of the method will be lost if an ordinary burette is used without special precautions. The titration flask should be kept stoppered as much as possible during titrations.

The reagent is standardized as follows:-

1. Place 10 to 15 ml of substantially dry methanol in the dry, stoppered 125 ml Erlenmeyer flask. Titrate the methanol with the reagent until the brown, iodine colour persists. Shake vigorously near the end of the titration and continue titrating until the colour change is permanent. The colour change from yellow to brown is quite sharp and easily reproducible. If the flask is not stoppered as much as possible, atmospheric moisture may cause the end point to fade.
2. Weigh out 0.05 to 0.08 grams of distilled water into the flask (2 to 4 drops from a weight pipette or 1 or 2 drops from a weighted medicine dropper and bottle). Titrate to the same end point as (1) above.

A convenient arrangement for weighing the distilled water and Hydrolube samples is to use a 1 or 2 ounce bottle fitted with a medicine dropper. Weights are obtained by difference. One drop will normally weigh between 0.05 and 0.06 grams.

3. Make several check determinations. The reagent should be checked each time it is used.

To determine the amount of water in the sample:-

1. Place 10 to 15 ml of substantially dry methanol in the dry, stoppered 125 ml Erlenmeyer flask and titrate to the brown end point as (1) above.
2. Add a weighed quantity (5 to 10 drops) of the representative hydrolube and again titrate to the same end point.
3. Calculate the per cent as follows:-

Standardization: $$\text{Factor} = \frac{\text{grams of water}}{\text{ml of reagent}} \times 100$$

Analysis: $$\text{Water, weight \%} = \frac{\text{ml of reagent}}{\text{grams of sample}} \times \text{Factor}$$

Viscosity Index Extension (VI_E)

A new formula for calculating Viscosity Indices above 100 is based on the recommendations of the ASTM in Method D.2270-64. This goes a long way towards removing the anomalies in the significance of V.I. for high viscosity index oils. The effect of the new method is shown in Fig.10.15 where curve B shows the results obtained by plotting the log of the viscosity at 100 degrees F against VI. This curve is an asymptote which for an oil having a viscosity of 10 centistokes at 210 degrees F never exceeds approximately 208 VI, even if the viscosity at 100 degrees F becomes equal to that at 210 degrees F. Curve A shows VI_E plotted against the log of viscosity at100 degrees F

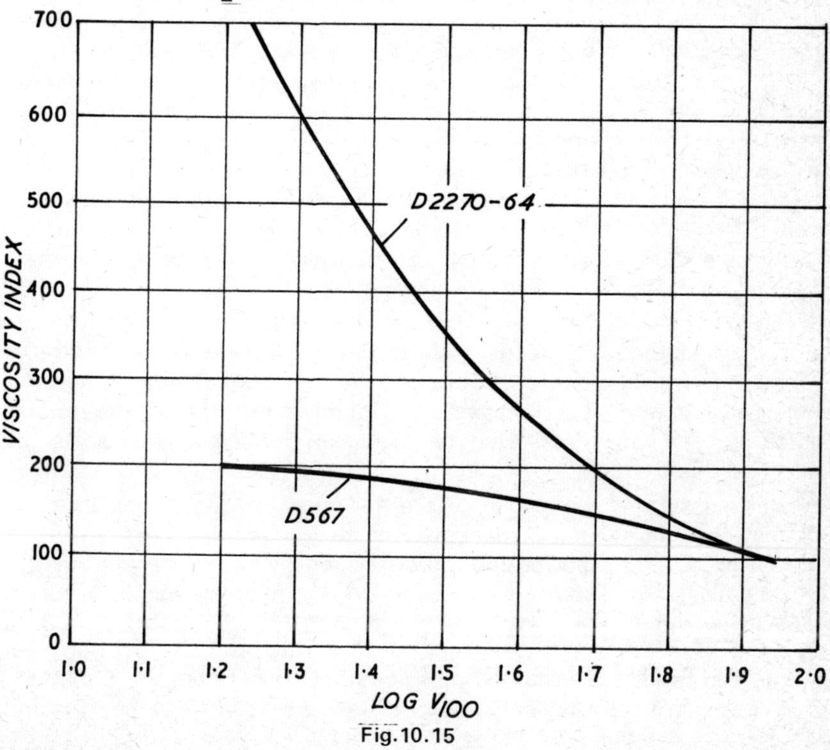

Fig.10.15

for an oil having a viscosity of 10 centistokes at 210 degrees F. This curve will be seen asymptotic to the vertical axis, i.e. in the case of an ideal oil with similar viscosities at 100 degrees F and 210degrees F would yield a VI_E of infinity.

The method is still not exact. Thus Fig. 10.16 shows the region in which anomalies still occur, this being a series of plots for four different oils having the same VI by the original method, plotted against viscosity at 210 degrees F on one axis, and against VI on the other. However, the relationship between VI and VI_E is reasonably consistent for oils having a viscosity above 4 centistokes at 210 degrees F, and thus such anomalies are largely confined to the low viscosity ranges outside the normal application of hydraulic fluids.

Wear Tests

A variety of empirical test methods have been developed for measurement of the lubricating properties of oils, probably the best known being the 'ball' tests. A typical example is the 4-ball test. Briefly, this consists of spinning a ½ in diameter steel ball clamped in a chuck on a constant speed motor in the cavity formed by three other ½ in diameter balls squeezed together in a clamping ring. The assembly is maintained in position in a bowl filled with the oil to be tested. The bowl itself is free to rotate on a thrust bearing and a vertical load is applied to the top ball by means of pressure exerted by the underlying balls being forced upwards by means of weights hung on a counterbalanced lever. During rotation this load causes a turning moment to be applied to the bowl which can be measured by means of a spring and recording drum actuated by a calibrating arm attached to the bowl.

The performance of the lubricant can thus be examined under several predetermined loads, whilst an indication of its extreme pressure properties can be obtained from the load at which seizure becomes apparent – as shown by violent movement of the torque indicator. The balls can also be examined for wear after a run of specified duration.

Such tests are mainly of value for assessing the qualities of conventional lubricants. A generally more satisfactory method for the evaluation of the lubricating qualities of hydraulic fluids is the Vickers Test for 'Maximum Severity Oils'. This employs a Vickers vane pump, type 105C, which is run at a speed of 1200 rpm, a fluid pressure of 2000 psi (140 kg/cm^2) with a flow rate of 5 gallons per minute.The fluid is maintained at a constant temperature of 65 degrees C, simulating conditions of severe overload. The quality of the fluid can then be assessed on the wear (measured as loss in weight) on the

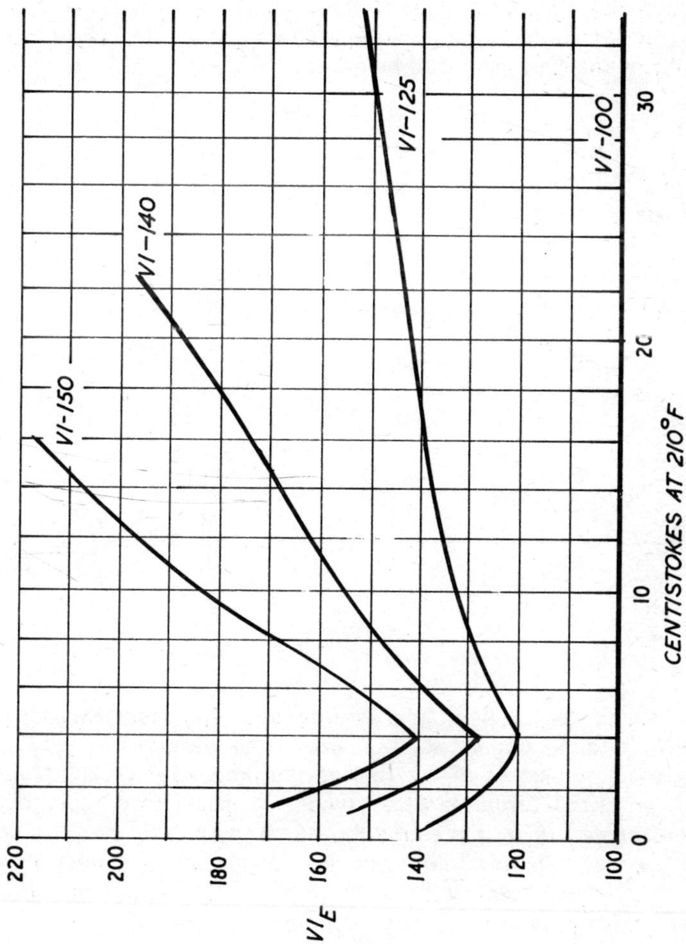
VI-150
VI-140
VI-125
VI-100
220
200
180
160
140
120
100
0
10
20
30
CENTISTOKES AT 210°F
VI_E

Fig. 10.16

inner and outer port plates, cam ring and vanes of the pump, after a specified duration of test. Test duration may be specified in hours requirement, or cease when excessive noise develops in the pump, or there is obvious loss of suction, due to excessive wear. Results are comparative – e.g. see Fig.10.17 - an individual test normally being discontinued when the rate of wear becomes very steep, or noise becomes excessive (indicating excessive wear, which is subsequently measured). In addition to actual wear measurements, the condition of the pump after the test may also be given.

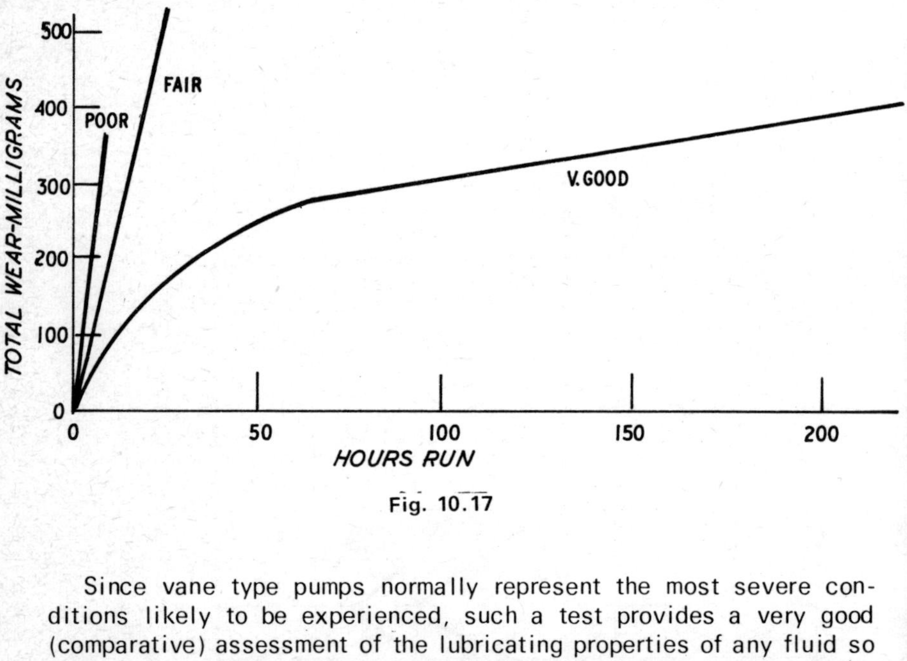

Fig. 10.17

Since vane type pumps normally represent the most severe conditions likely to be experienced, such a test provides a very good (comparative) assessment of the lubricating properties of any fluid so tested. In practical circuits, other types of pumps (e.g. gear pumps and piston pumps) do not present the same 'maximum severity' conditions. However, similar tests may be conducted with gear rigs to evaluate the likely wear with gear pumps. The standard machine is the IAE gear rig, as per I.P. Test Procedure No.166/6OT.

APPENDIX 1: FLUID PARAMETERS

Density and specific weight

Specifically, density is the **mass** of a fluid per unit volume; and specific weight is the **weight** per unit volume.

$$\text{Thus: density, } \rho = \frac{\text{weight}}{\text{volume x g}} \text{ e.g. } = \frac{\text{lbm}}{\text{Vol}} = \frac{\text{lbf}}{\text{Vol x g}}$$

$$\text{specific weight, w} = \frac{\text{weight}}{\text{volume}} \text{ e.g. } = \frac{\text{lbf}}{\text{vol}}$$

For engineering calculations specific weight is normally implied, standard units being:—

pounds per cubic foot
pounds per cubic inch
kilograms per litre

The change of density (or specific weight) with temperature and pressure may be rendered in terms of relative density, i.e. the density at a given temperature (or pressure) relative to the density at standard temperature and pressure.

Specific Gravity (SG)

Specific gravity is a dimensionless quantity and is simply the ratio of the density (or specific weight) of a fluid to the density (or specific weight) of water. In the case of equations for engineering calculations it is often desirable to render density or specific weight, where it appears as a factor, in terms of specific gravity, thus avoiding any possible confusion between the true numerical values density and specific weight, which are to be employed in the formula.

The significance of specific gravity as a hydraulic fluid parameter is that it gives an indication of the weight of the fluid in the system, or more directly a comparison of fluid weights for a given system where different fluids may be considered. Also the higher the specific gravity of the fluid the more difficult the fluid is to lift in the suction part of the system, which may need particular care in the design of the suction side in order to avoid the possibility of cavitation and erratic pump operation.

Viscosity

The viscosity of a fluid is a measure of its internal resistance. Dynamic viscosity is defined in terms of the force in dyne between two parallel laminae or layers of fluid each 1 sq.cm. in area with a slip velocity of 1 cm/sec between them, the corresponding unit of viscosity being the **poise**. Because the dynamic viscosity of real fluids determined in poises is invariably a fractional quantity, the more unusual unit employed for expressing dynamic viscosity is the **centipoise**, or one hundredth of a poise.

For engineering calculations it is usually more convenient to employ the **kinematic viscosity** rather than dynamic viscosity, this being determined as the absolute dynamic viscosity divided by the mass density of the fluid. The standard unit is the **stoke** (cm^2/sec.) but for the same reason as above the practical unit is invariably taken as one hundredth of a stoke or **centistoke.**

The centistoke is the standard unit commonly employed in both English and metric formulas. If English units are employed throughout, however, the corresponding unit of kinematic viscosity is either ft^2/sec, or $inch^2$/sec, neither of these units having a name.

1 centistoke = 0.00001076 ft^2/sec.
1.0 ft^2/sec = 93000 centistokes
1 centistoke = 0.00155 in^2/sec.
one in^2/sec. = 648 centistokes

In practice fluid viscosities are determined by viscometer measurement, these instruments being calibrated in arbitrary scales. The three scales employed for measurement of the viscosity of oils and hydraulic fluids are: –

Redwood No. 1 seconds, in Britain (normally quoted at 70°F, 140°F & 200°F).

Engler degrees, in Continental Europe.
Saybolt Universal Seconds (SUS or SSU) in America, (normally quoted at 100°F and 210°F)

There are no exact conversions between the various arbitrary scales, nor for conversion of arbitrary viscosities in seconds or degrees to kinematic viscosities in centistokes, or ft^2/sec. (or in^2/sec) units. Close approximate conversions can be made by reference to conversion scales of conversion tables. It should be noted that such conversions apply only at the same temperature as the original measurement.

In the case of a non-Newtonian fluid the instantaneous viscosity of the fluid is dependent on the shear stress in that fluid at that particular moment. If necessary, a specific viscosity figure can be obtained with a viscometer which ensures a uniform shear rate throughout measurement. Such a figure will, however, have limited practical value, unless the shear stability characteristics of the fluid are also known.

Variations of viscosity with temperature is one of the most significant parameters with hydraulic fluids, affecting both the performance and selection of a fluid. This can be fully expressed by plotting a characteristic curve for the fluid on an ASTM chart – Fig. A. The scales of an ASTM chart are so designed that the characteristic curve for most fluids in linear. Given a number of spot readings only for viscosity and temperature, a close approximation to the viscosity temperature characteristics of that fluid at intermediate temperatures can be obtained by joining these points with a straight line; and for temperatures outside the range covered by the spot values, by extending this line in either direction.

Viscosity-pressure characteristics are summarised in Chapter 2.

Surface Tension

Surface tension may be significant as a fluid property as affecting:–

(i) Foaming characteristics at the free liquid surface or interface between two non-miscible fluids.
(ii) The ability of the fluid to 'wet' a metal surface.
(iii) Inherent leakage past seals and at joints, etc.

Items (i) and (ii) can be adequately controlled by additives, if necessary. Item (iii) is not normally significant in the case of oil fluids which have adequate surface tension to make sealing relatively easy. Fluids which have a surface tension of less than 30 $dynes/cm^2$, however, are troublesome to seal, needing particular attention to joints and often having to accept that some leakage will be inevitable with practical designs of seals.

TABLE A.1

VOLUME CORRECTION COEFFICIENTS

Specific Gravity 60/60 °F	k_v per °F	k_v per °C
.5994 - .6022	.00096	.00173
.6023 - .6053	.00095	.00171
.6054 - .6082	.00094	.00169
.6083 - .6112	.00093	.00167
.6113 - .6144	.00092	.00166
.6145 - .6176	.00091	.00164
.6177 - .6211	.00090	.00162
.6212 - .6239	.00089	.00160
.6240 - .6269	.00088	.00159
.6270 - .6299	.00087	.00157
.6300 - .6336	.00086	.00155
.6337 - .6368	.00085	.00153
.6369 - .6401	.00084	.00151
.6402 - .6433	.00083	.00149
.6434 - .6466	.00082	.00148
.6467 - .6506	.00081	.00146
.6507 - .6539	.00080	.00144
.6540 - .6582	.00079	.00142
.6583 - .6634	.00078	.00140
.6635 - .6677	.00077	.00139
.6678 - .6719	.00076	.00137
.6720 - .6765	.00075	.00135
.6766 - .6810	.00074	.00133
.6811 - .6849	.00073	.00131
.6850 - .6902	.00072	.00130
.6903 - .6958	.00071	.00128
.6959 - .7007	.00070	.00126
.7008 - .7054	.00069	.00124
.7055 - .7108	.00068	.00122
.7109 - .7161	.00067	.00121
.7162 - .7212	.00066	.00119
.7213 - .7262	.00065	.00117
.7263 - .7315	.00064	.00115
.7316 - .7366	.00063	.00113
.7367 - .7417	.00062	.00112
.7418 - .7471	.00061	.00110
.7472 - .7523	.00060	.00108
.7524 - .7574	.00059	.00106
.7575 - .7621	.00058	.00104
.7622 - .7672	.00057	.00103
.7673 - .7725	.00056	.00101
.7726 - .7777	.00055	.00099
.7778 - .7833	.00054	.00097
.7834 - .7894	.00053	.00095
.7895 - .7949	.00052	.00094
.7950 - .8006	.00051	.00092
.8007 - .8074	.00050	.00090
.8075 - .8158	.00049	.00088
.8159 - .8244	.00048	.00086
.8245 - .8329	.00047	.00085
.8330 - .8424	.00046	.00083
.8425 - .8519	.00045	.00081
.8520 - .8626	.00044	.00079
.8627 - .8748	.00043	.00077
.8749 - .8909	.00042	.00076
.8910 - .9079	.00041	.00074
.9080 - .9305	.00040	.00072
.9306 - .9523	.00039	.00070
.9524 - .9755	.00038	.00068
.9756 - 1.0000	.00037	.00067
1.0001 - 1.0184	.00036	.00065
1.0185 - 1.0375	.00035	.00063
1.0376 - 1.0552	.00034	.00061
1.0553 - 1.0759	.00033	.00059
1.0760 - 1.1249	.00032	.00058

Volume coefficient k_v multiplied by temperature change gives the effective volume change (contraction or expansion) with that change of temperature for typical mineral oils.

Vapour Pressure

The vapour pressure of a fluid is the pressure exterted by the saturated vapour in contact with the surface of the fluid at a specified temperature. The higher the vapour pressure the more volatile the fluid, and/or the nearer it is to boiling point. Fluids with a high vapour pressure, either due to their volatile nature, or because of a high operating temperature, are therefore prone to 'flash' into vapour under suction conditions, thus setting a specific limit to the net positive suction head a pump can accommodate without cavitating.

Additionally, as the boiling point of a fluid is approached,the more volatile fractions will come off first, progressively changing the nature of the fluid. This is seldom significant in oil fluids at normal working temperatures, but with water fluids progressive loss of water may be experienced at quite moderate working temperatures.

Toxicity and Hazard

Fluids are classified as hazardous if they, or their vapour, are flammable, toxic, or chemically reactive. This applies particularly to fluid, or vapour, which has escaped by leakage. The hazard involved is normally determined and classified under three separate headings:—

(i) danger to health
(ii) flammability
(iii) reactivity.

Industrial fluids are given hazard ratings from 0 (non-existent) to 4 (very hazardous) in each rating. This method is not applied to hydraulic fluids, any hazard present being described in more general terms (e.g. oil fluids are flammable); except that reactivity is normally expressed explicitly as reservations of compatibility.

Cloud Point

The cloud point of a mineral oil is that temperature at which waxes or other solids normally present in solution tend to crystallise out, or come out of solution. This can lead to clogging or partial choking of the system, if operating at the cloud point of the fluid.

Pour Point

This is the temperature at which the thickening action of separation of the waxy constituents is so marked that the fluid ceases to flow. On the ASTM curve this would be marked by an abrupt rise in viscosity. The pour point (temperature) is determined with regard to specific flow conditions — see Chapter II.

Thermal Expansion

The coefficient of volumetric expansion of an oil remains practically constant over the usual range of working temperatures encountered in hydraulic systems. Specific values are largely related to the specific

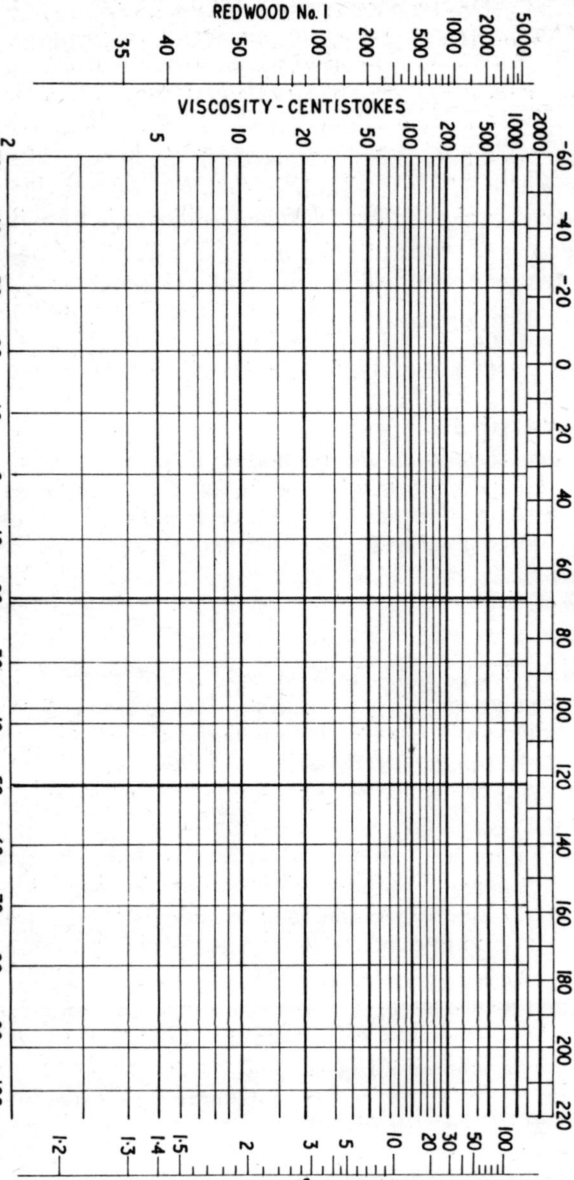
REDWOOD No. 1
35 40 50 100 200 500 1000 2000 5000
VISCOSITY - CENTISTOKES
2 5 10 20 50 100 200 500 1000 2000
TEMPERATURE °C
-50 -40 -30 -20 -10 0 10 20 30 40 50 60 70 80 90 100
TEMPERATURE °F
-60 -40 -20 0 20 40 60 80 100 120 140 160 180 200 220
VISCOSITY °E
1·2 1·3 1·4 1·5 2 3 5 10 20 30 50 100

Fig. A

gravity of the oil – see Table A.1. The coefficient of volumetric expansion, however, decreases appreciably with increasing pressure, e.g. of the order of 0.000025 per degrees C per 1000 psi. The relative volume of a pressurised fluid is thus less than that predicted on the basic of volume correction coefficients alone - see also Chapter 2.

Specific Heat

The specific heat of a fluid is defined as the ratio of the heat required to raise the temperature of a given volume of fluid by one degree to that required to raise the temperature of the same volume of water by the same amount. It can thus be specified either in B.Th.U or calory-gm-cm units. The specific heat is not constant but varies with temperature, although for practical calculation a constant value is often assumed (based on the specific heat over the 62–63 degrees F temperature range).

In the absence of specific figures the following semi-empirical formula can be used to calculate the specific heat of hydraulic oils.

$$\text{specific heat} = \frac{0.388 \times 0.00045t}{\sqrt{SG}} \text{ BthU/lb per degree F}$$

where t = temperature in degrees F
SG = specific gravity of the fluid at 60 degrees F

The specific heat of a hydraulic oil may be appreciably modified by the presence of additives, and also of contaminants in the fluid. See also Table A.2.

TABLE A.2
SPECIFIC HEAT OF FLUIDS*

Fluid	Specific Heat
Water	1.0
Mineral Oil (DTD 585)	0.456
Water-Glycol at 25°C	0.718
at 60°C	0.774
at 80°C	0.785
Phosphate Ester at 15°C	0.43
Silicone	0.39

*See Chapters 7 & 8 for specific properties of proprietary fluids.

Thermal Conductivity

The thermal conductivity is a measure of the ability of a fluid to dissipate heat. In a practical system heat dissipation may be hindered by the formation of boundary layer films and thus fluids which do not 'wet' the internal surfaces tend to have lower thermal conductivities. The lower the thermal conductivity of the fluid the higher the working temperature of the fluid will tend to be, under similar operating conditions. Thus mineral oils, having a generally low thermal conductivity, will tend to run at higher working temperatures than water-base or water-glycol fluids used in a similar system.

The thermal conductivity of a hydraulic oil can be estimated from the following semi-empirical formula

$$\text{Thermal conductivity} = \frac{0.821 - 0.000244t}{SG} \text{ BthU/sq ft/hr/degree F/inch}$$

where t = temperature in degrees F
SG = specific gravity of fluid at 60 degrees F.

By comparison the thermal conductivity of water is some 4.5 times greater, and that of water-glycol solutions some 3 times greater. See also Table A.3.

TABLE A.3.
THERMAL CONDUCTIVITY OF FLUIDS*
(calories /cu/sec/°C)

Fluid	Thermal Conductivity $\times 10^{-4}$
Water at 10°C	14.7
50°C	15.4
80°C	16.0
Mineral Oil (DTD 585)	3.24
Water-Glycol	9.61
Phosphate Ester	5.30
Glycerine at 70°C	6.8

*See Chapters 7 & 8 for specific properties of proprietary fluids.

Entropy (u)

If work is added to a fluid in a closed system with no heat transfer to or from the system, the amount of work added depends only on the end states of the process and the energy content consists of kinetic energy, potential energy and internal energy, (although one or more of these may be negligible). This energy content is called entropy (u) and the relationship expressed is that of the first law of thermodynamics. If a change in entropy is involved, then this represents a heat flow, into or out of, the system.

Enthalpy (h)

When heat is applied to a fluid a change in internal energy results, the increase representing the difference between the heat supplied and the work done by the fluid. In an adiabatic process, where no transfer of heat in involved, work is done by the fluid entirely at the expense of its internal energy. This may be better expressed in terms of total heat, or enthalpy, when enthalpy is defined as the sum of the internal energy and the product of pressure and specific volume, viz

$$h = u + PV$$

or, total heat $= E = \frac{P}{Jw}$

where E = internal energy, expressed in heat units
J = Joules mechanical equivalent of heat
P/w = flow energy

In general, engineering hydraulic calculations are not concerned with thermodynamic problems, although these may arise in specific cases.

Colour

The natural colour of a hydraulic fluid normally has no significance. Thus mineral oil blends may have a natural colour, depending primarily on their origin, or be clarified by refining processes (e.g. particularly to remove darker colouration). A limited number of fluids may have a characteristic colour (e.g. milky-white for an oil-in-water emulsion, and for some water-glycol solutions); but many may have colour added as a means of identification. Some identifying colours are summarised in Table A.4.

TABLE A.4

COLOURS ADOPTED FOR HYDRAULIC FLUIDS

Fluid	Type	Colour
DTD 585	Mineral Oil (Aircraft)	Deep Red
DTD 391 (Obsolete)	Castor Base (Aircraft)	Deep Blue
DTD 641 (Obsolete)	Castor Base (Aircraft)	Yellow-Brown
DTD 900/4081	Castor Base (Aircraft)	Pale Yellow
Skydrol 500	Phosphate Ester	Purple
Skydrol 7000	Phosphate Ester	Green

Viscosity Gravity Constant (VGC)

This is an empirical relationship based on specific gravity and viscosity which may be determined as an approximate indication of the aromatic content of an oil or blend of oils. Its significance is similar to the Aniline point in that the VGC is an indication of the likely degree of swelling of synthetic elastomers immersed in the fluid. In practice, the VGC gives unreliable predictions.

Aniline Point

The aniline point of a mineral oil is the lowest temperature at which the oil is completely miscible with an equal volume of freshly distilled aniline. It is a general indication of the aromatic content of the oil (paraffinic oils having a high aniline point and aromatic oils a low aniline point); and because of this is sometimes used as a form of compatibility index – see Chapter 9.

Flash Point

The flash point broadly defines the relative fire hazard of a fluid. The flash point may also serve as an indication of the type of an oil or blend, since the more volatile the oil the lower the flash point, and vice versa. The flash point may be determined by various standard 'open' or 'closed' cup tests – see Chapter 10.

SAE Number

The SAE Number is an arbitrary classification of oils by viscosity, normally applied only to crankcase oils and gear oils – see Table A.5.

TABLE A.5
S.A.E. No. SPECIFICATION

	S.A.E. Number	Viscosity (Saybolt Universal Seconds) At 0 degrees F. min.	max.	At 210 degrees F. min.	max.	Lower temp. limit for no channelling
Crankcase Oils	5W	–	4 000	–	–	–
	10W	6 000	12 000	–	–	–
		(minimum viscosity at 0 degrees F can be waived provided viscosity at 210 degrees F is not below 40 S.U.S.)				
	20W	12 000	48 000	–	–	–
		(minimum viscosity at 0 degrees F can be waived provided viscosity at 210 degrees F is not below 45 S.U.S.)				
	20	–	–	45	less than 58	–
	30	–	–	58	less than 70	–
	40	–	–	70	less than 85	–
	50	–	–	85	110	–
Gear Oils	75	–	15 000	–	–	–40 degrees F
	80	15,000	100 000	–	–	–20 degrees F
	90	–	–	75	120	–10 degrees F
	140	–	–	120	200	20 degrees F
	250	–	–	200	–	–

NOTE: S.A.E. numbers with the affix 'W' are designated on viscosity at 0 degrees F consistent with 'Winter' operation. Other S.A.E. numbers for crankcase oils are based on viscosity at 210 degrees F.

TABLE A.6

THE SIGNIFICANCE OF OIL PARAMETERS*

Property	Measure of	Significance or application(s)
Aniline point	Solvent power	Assessment of aromatics (approximate) Solvent action on organic finishes Swelling effect on rubbers Determination of cetane number (approx.)
A.P.I. gravity	Density (empirical)	Weight/volume relationships
Ash content	Incombustible constituents	Assessment of contamination Assessment of additives Identification of soaps (greases)
B.S. and W.	Sediment and water	Measure of solid impurities and aqueous solutions likely to separate out on standing
Carbon content.	Carbon residue on burning	Amount of 'soot' on burning, staining effect on metals (heat-treatment baths)
Colour		Rough guide as to type or blend
Cetane number	Ignition power	High cetane number (50 or over) equivalent to high thermal efficiency
Channelling temperature	Homogenity	Parting of lubricant likely, leaving areas uncovered
Cloud point	Separation of wax 'cloud'	Likely temperature for clogging of filters
Dielectric constant	Electrical resistance	Breakdown voltage determined as test characteristic
Diesel index	Ignition power	Same as Cetane number
Drop point	Change of state of grease	Suitability for high temperatures.
Film strength	Homogenuity	Assessment of lubrication under severe loading
Fire point	Burning temperature of vapour	Continuous burning sustained
Flash point	Burning temperature of vapour	Momentary ignition
Neutralisation number	Acidity or alkalinity	Assessment of deterioration or oxidation of used oils
Octane number	Anti-knock properties	High octane number designates higher thermal efficiencies utilising increased compression ratio
Penetration number	Measurement of hardness (greases)	Softness of greases
Pour-Point	'Freezing' point	Limiting minimum service temperature for ready flow
S.A.E. number	Arbitrary classification by viscosities	Identification of oil grade
Saponofication number	Fat and fatty acid content	As neutralisation number; also assessment of greases
Viscosity index	Viscosity/temperature relationship	High viscosity index for low change of viscosity with temperature
Viscosity	Fluid friction	Assessment of power losses, bearing heating, pump design requirements (hydraulic pumps)

* NOTE: only a limited number of these parameters are significant in the case of hydraulic fluids.

Neutralisation Number

The neutralisation number is a measure of the total quantity of mineral and organic acids in an oil, or the difference between the acid and the alkali in an oil. It can be used, broadly, to estimate the corrosion potential of an oil, and also for the estimation of the degree of oxidation present (by the change in the neutralisation number brought about by increasing acidity due to oxidation).

Sulphur Content

The sulphur content of an oil is mainly significant in internal combustion engine fuels since it leads to the formation of sulphurous and sulphuric acids in the presence of water (which is a product of combustion); and for oils used in contact with copper and copper alloys (which sulphur attacks). Since the latter metals are not normally used for hydraulic system components the sulphur content of oils has little significance for hydraulic work.

APPENDIX 2: PRESSURE DROP CHARTS

The following Charts for pressure drop in small bore pipes are calculated for a fluid specific gravity of 1.0. Pressure drop and flow rates are given in English and metric units on all Charts, but separate charts are given for inch-bore and metric-bore pipe sizes.

To correct the pressure drop for any other fluid specific gravity, simply multiply the chart value obtained by the actual specific gravity. For example, for a typical hydraulic oil of specific gravity 0.86, multiply chart values for pressure drop by 0.86.

Intermediate values of bore sizes can be interpolated.

VISCOSITY 10 CENTISTOKES

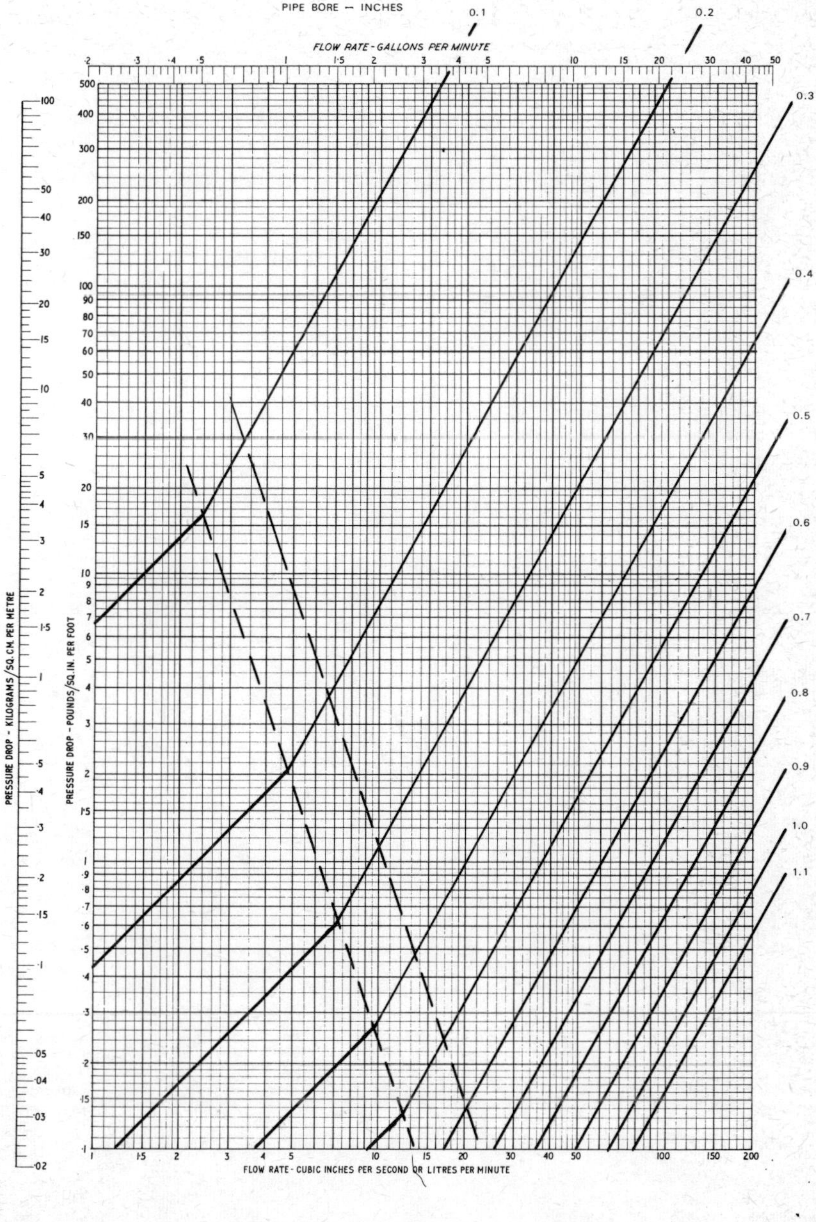

VISCOSITY 20 CENTISTOKES

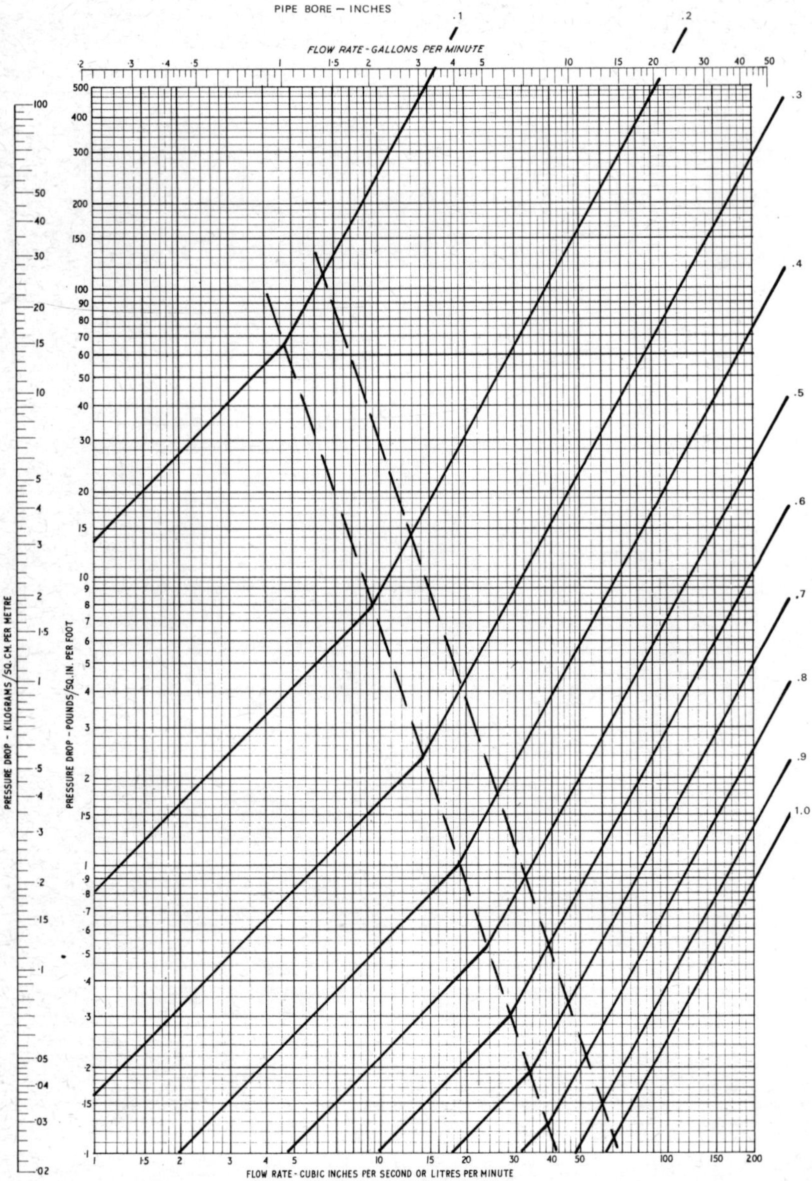

VISCOSITY 30 CENTISTOKES

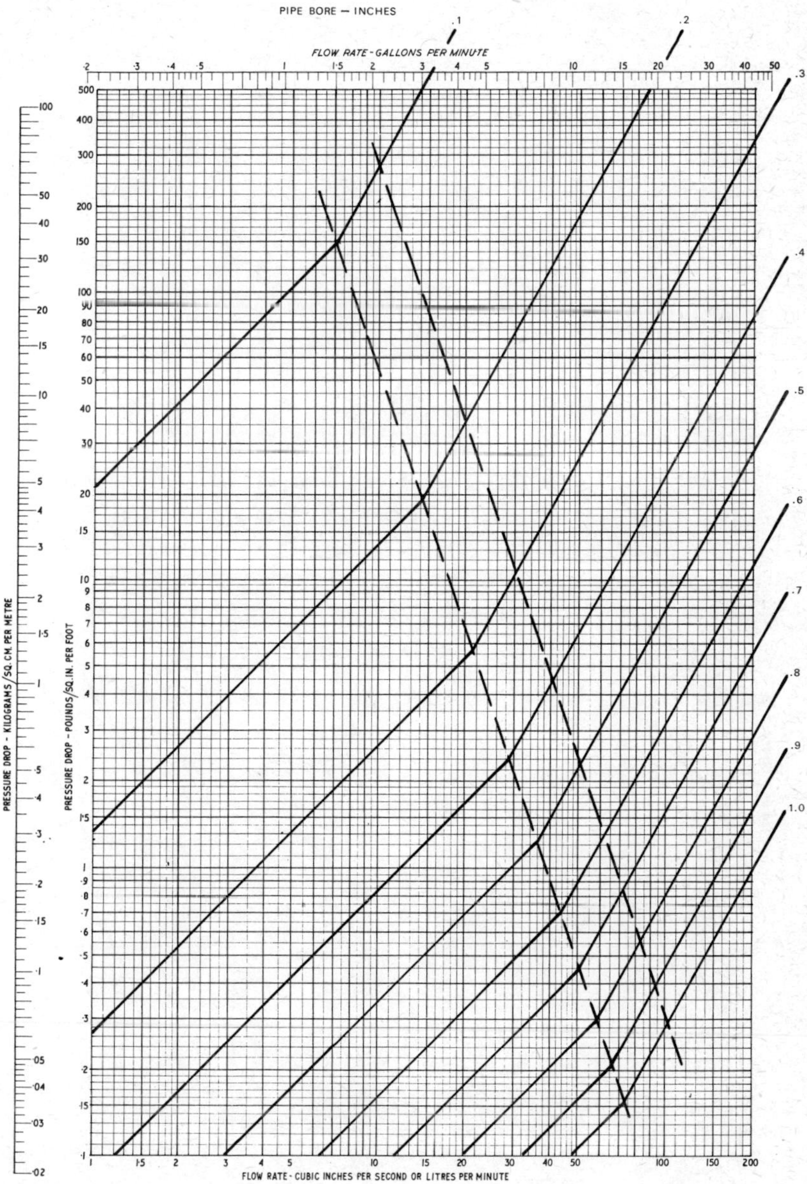

VISCOSITY 40 CENTISTOKES

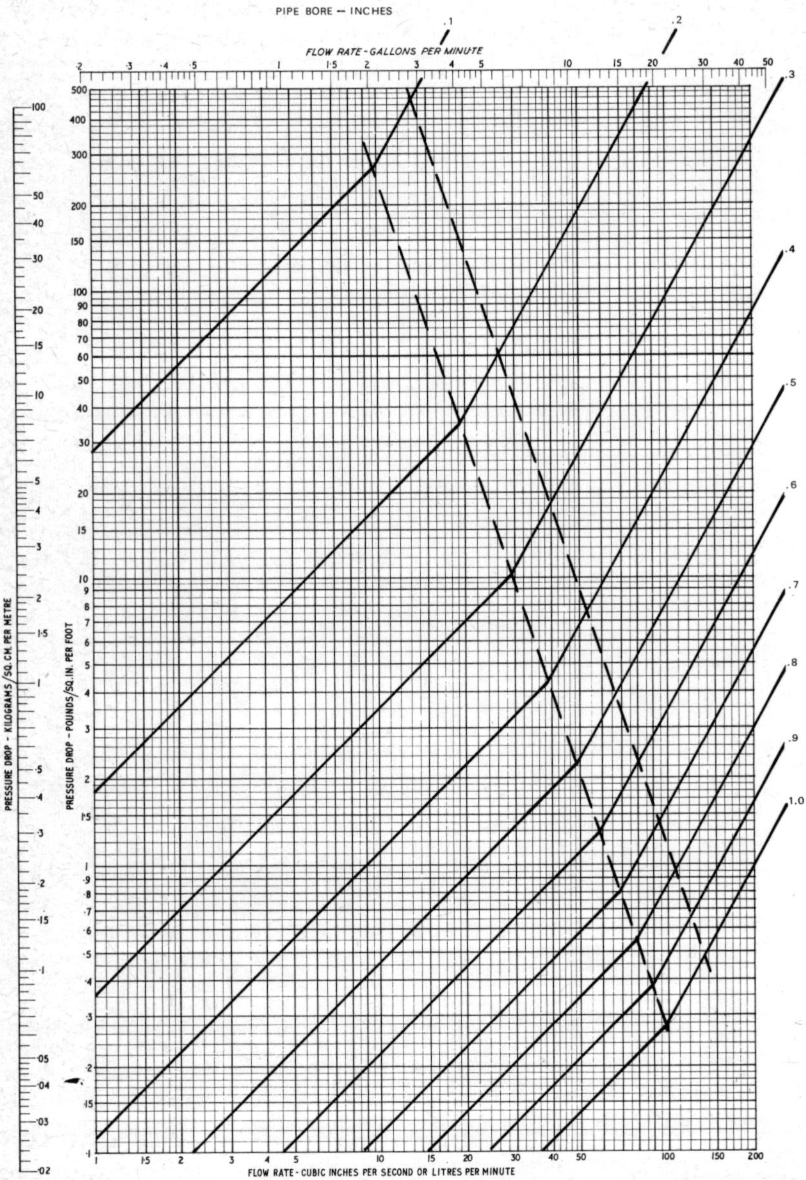

VISCOSITY 50 CENTISTOKES

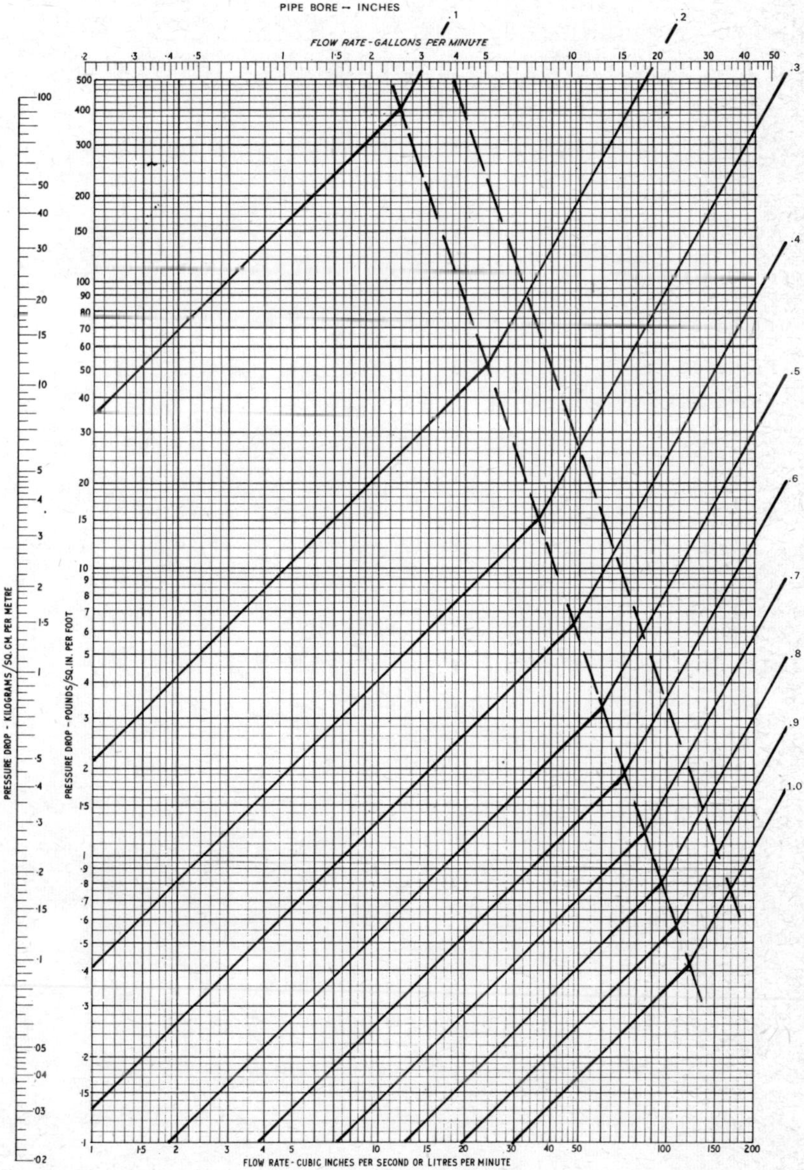

VISCOSITY 60 CENTISTOKES

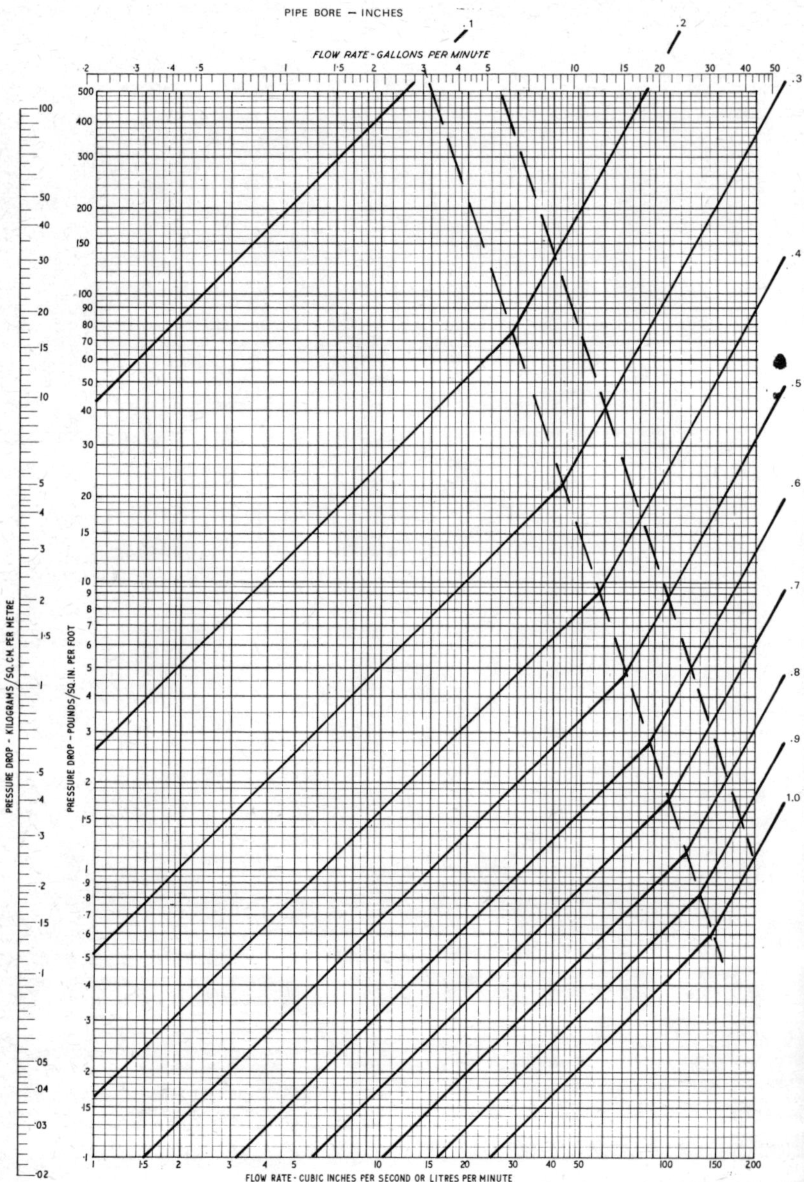

VISCOSITY 70 CENTISTOKES

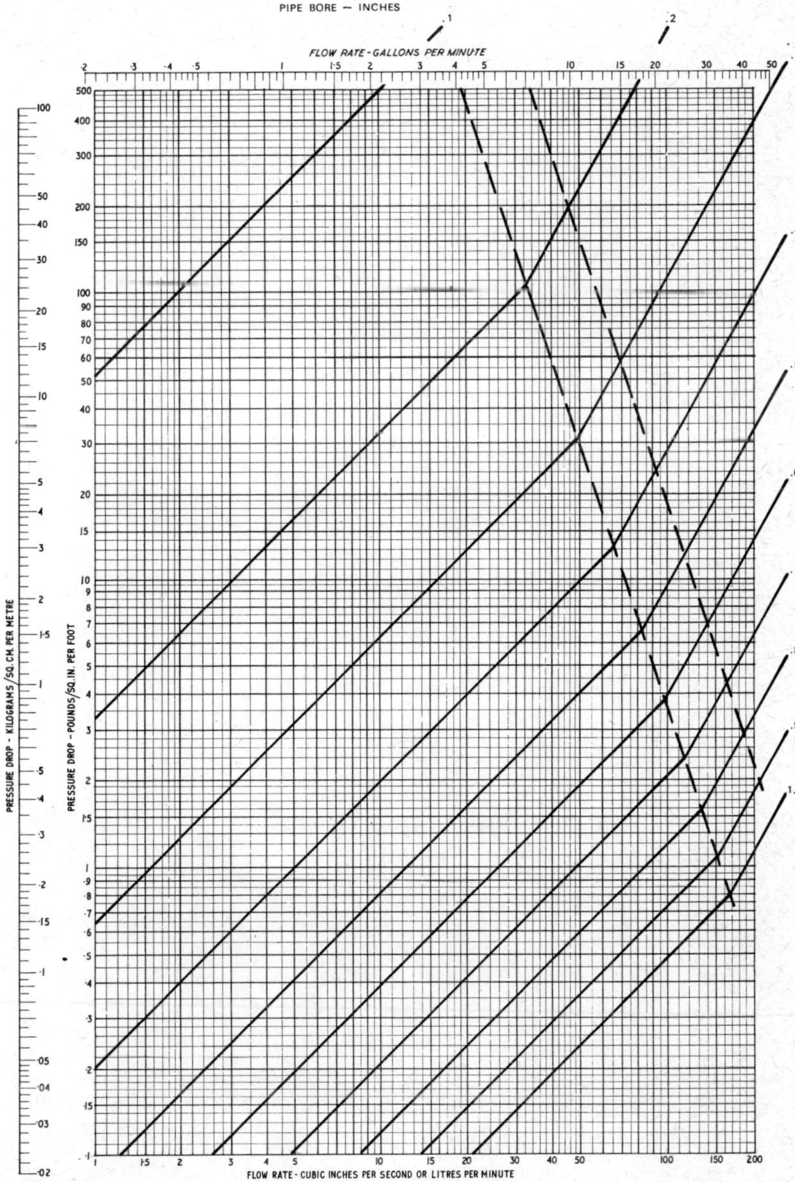

VISCOSITY 80 CENTISTOKES

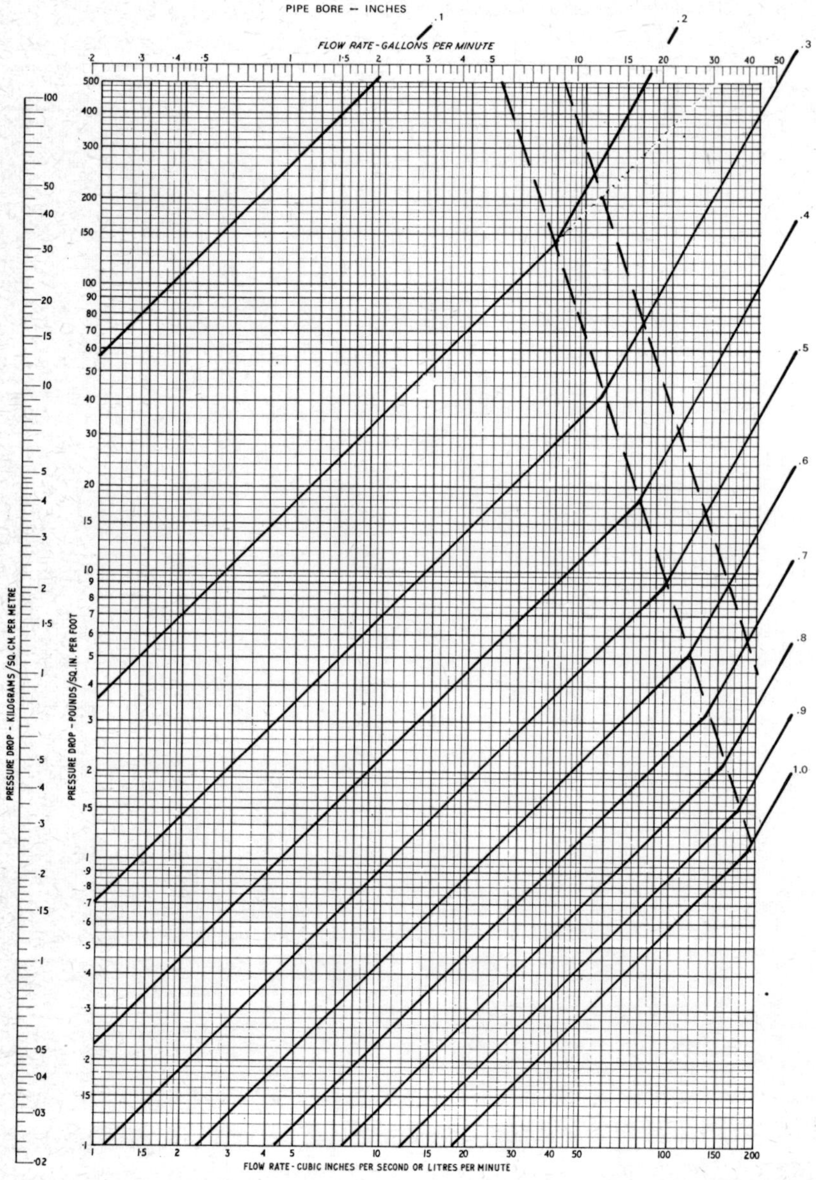

VISCOSITY 90 CENTISTOKES

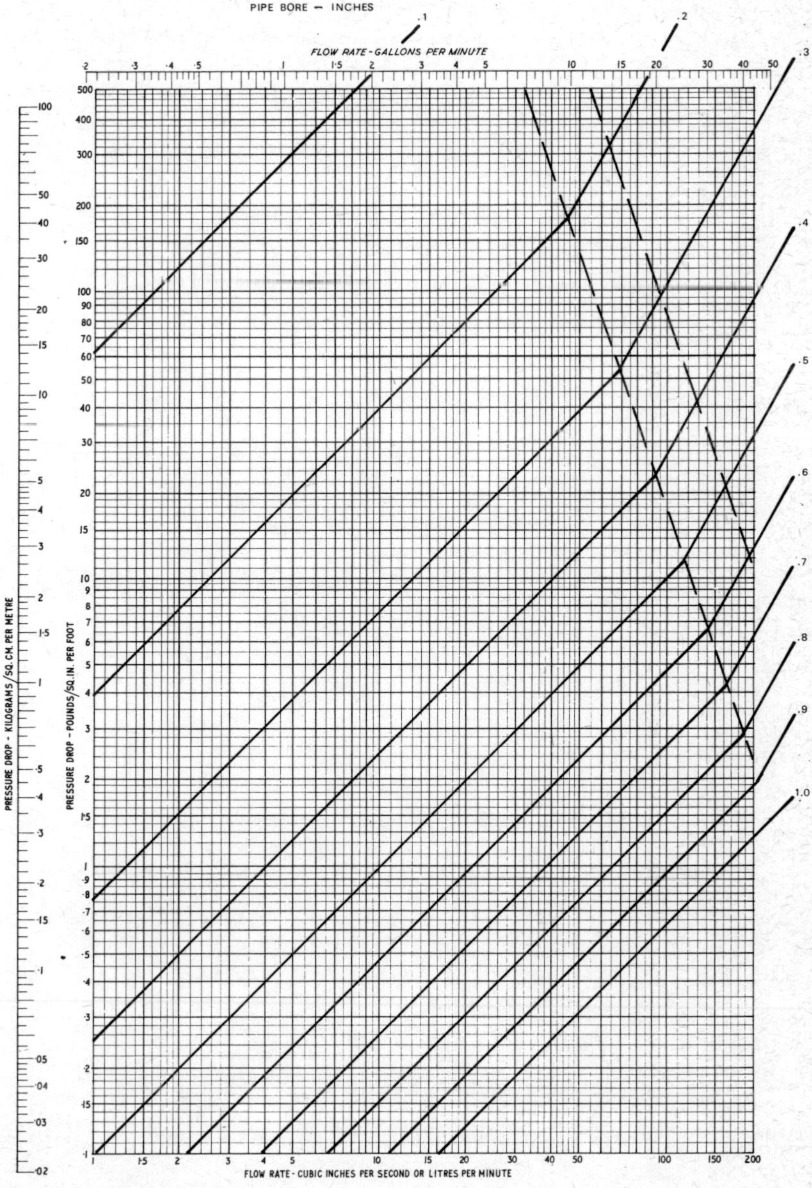

VISCOSITY 100 CENTISTOKES

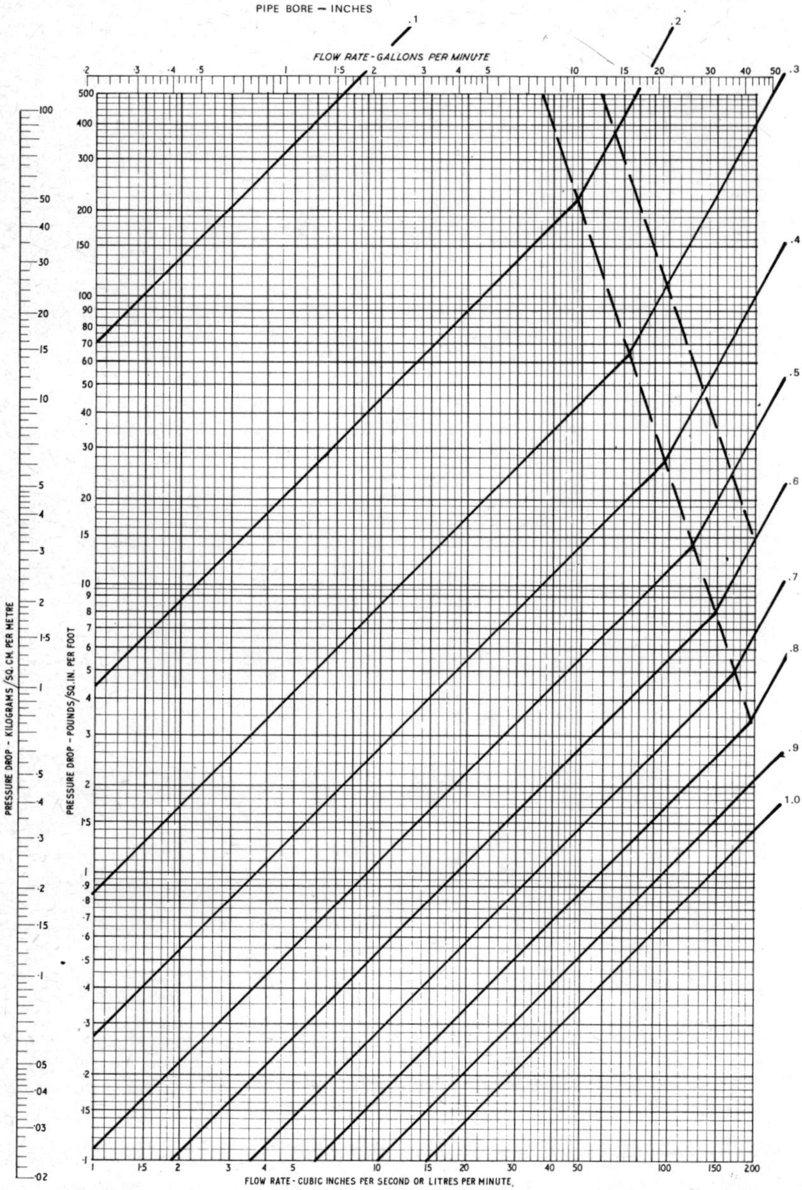

VISCOSITY 10 CENTISTOKES

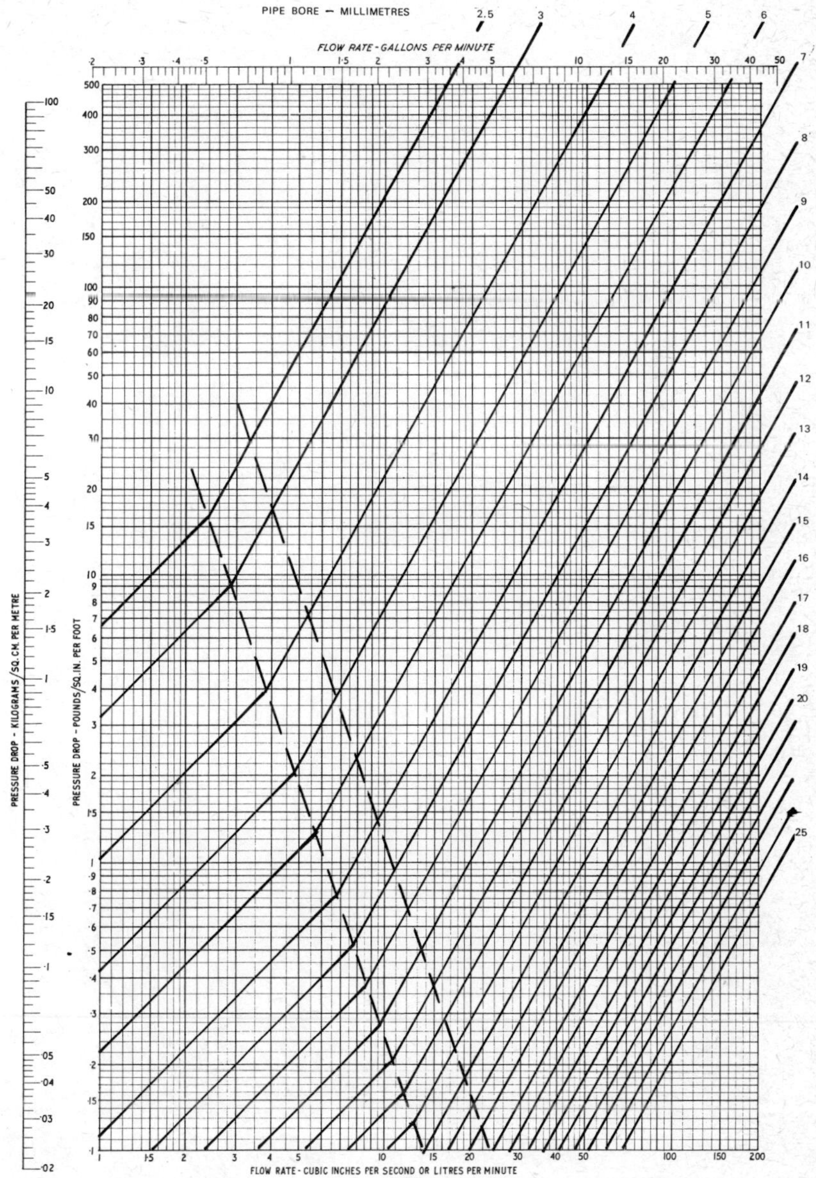

VISCOSITY 20 CENTISTOKES

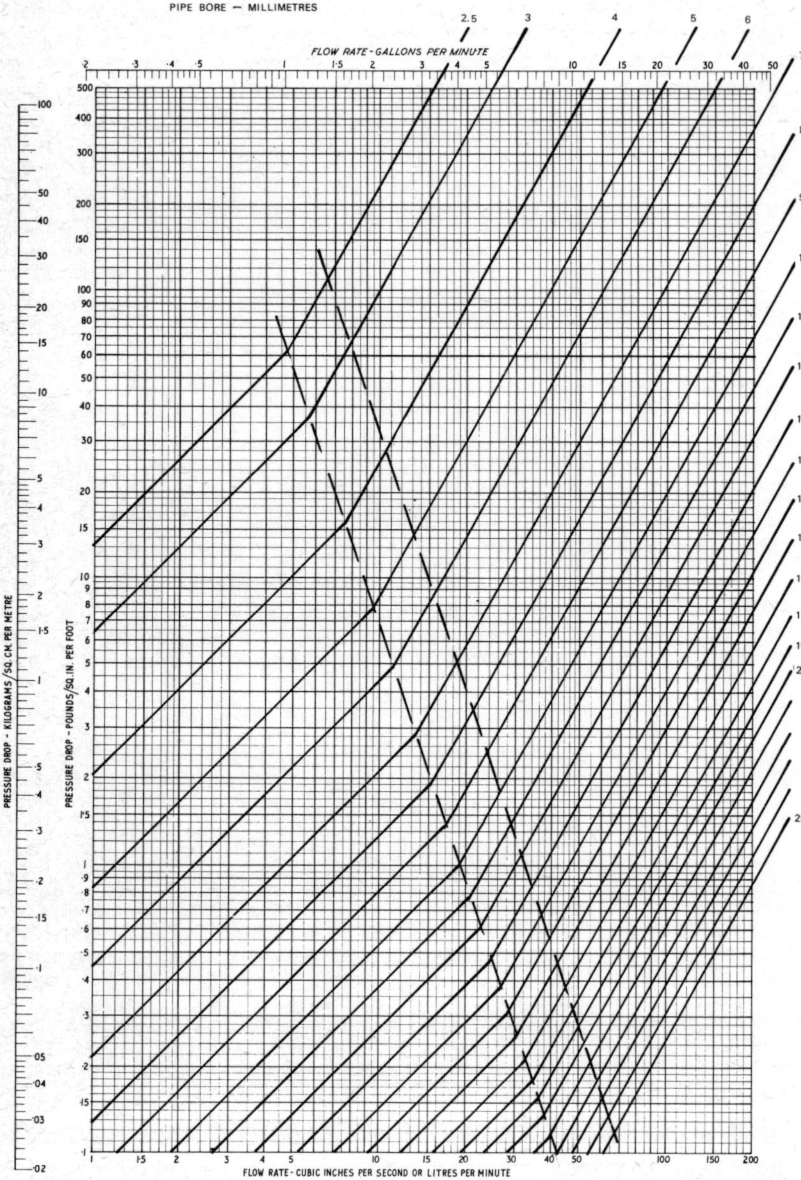

VISCOSITY 30 CENTISTOKES

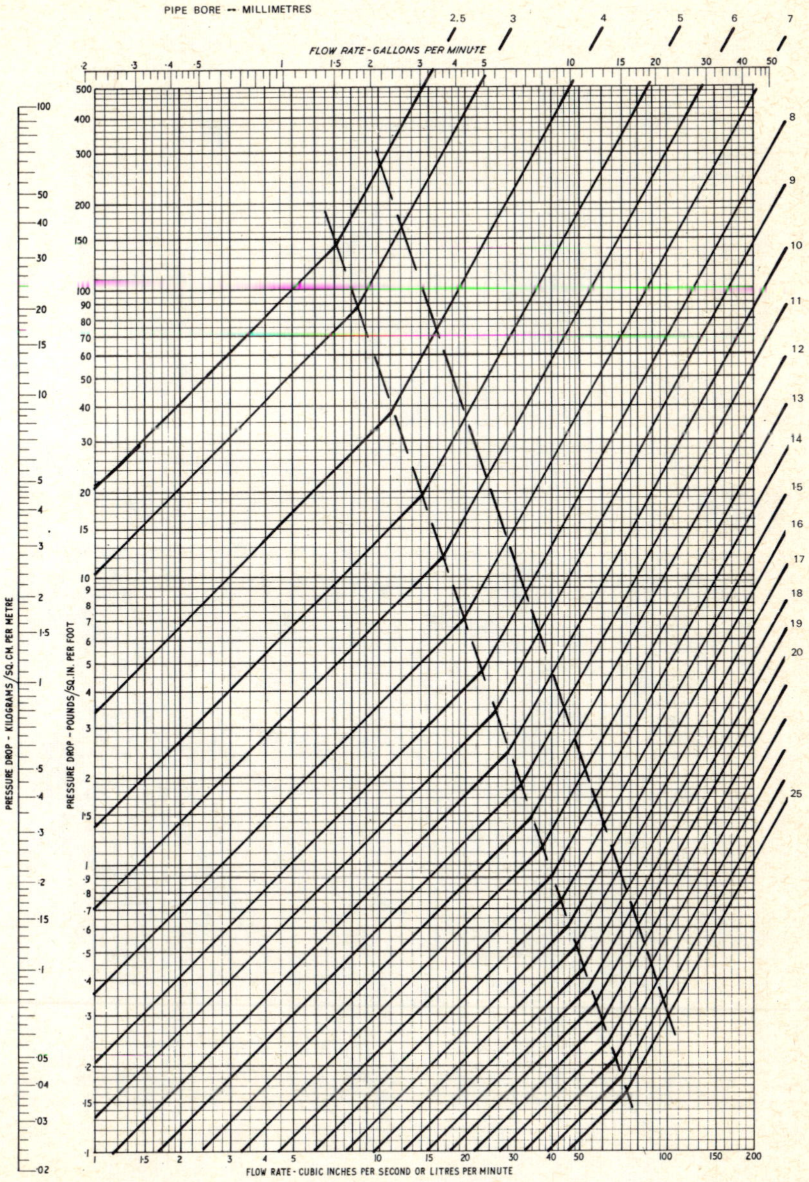

VISCOSITY 40 CENTISTOKES

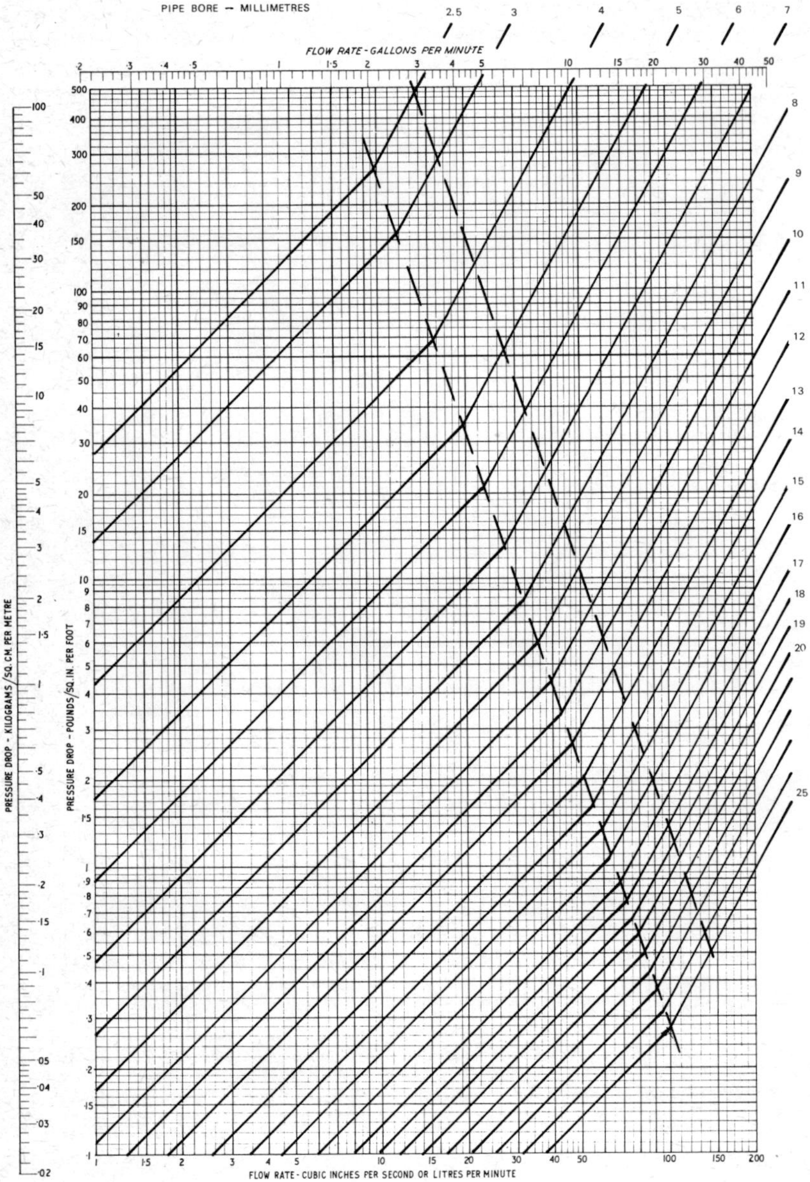

VISCOSITY 50 CENTISTOKES

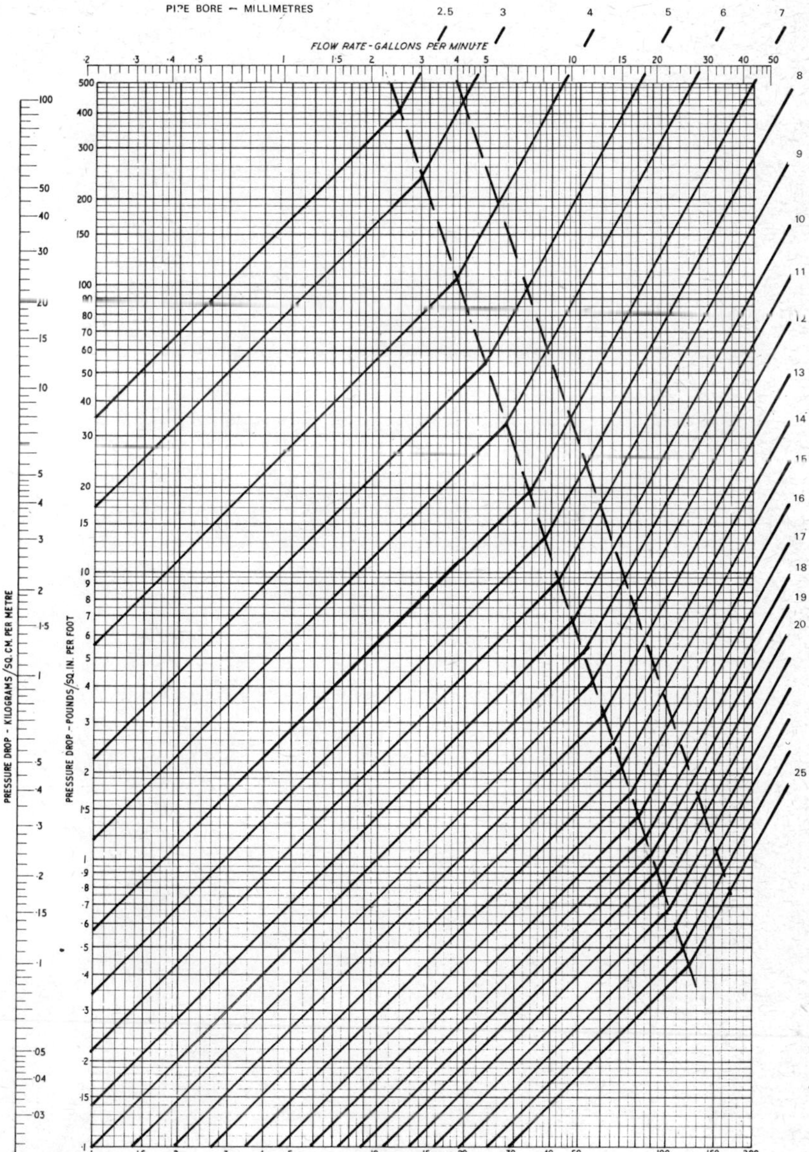

VISCOSITY 60 CENTISTOKES

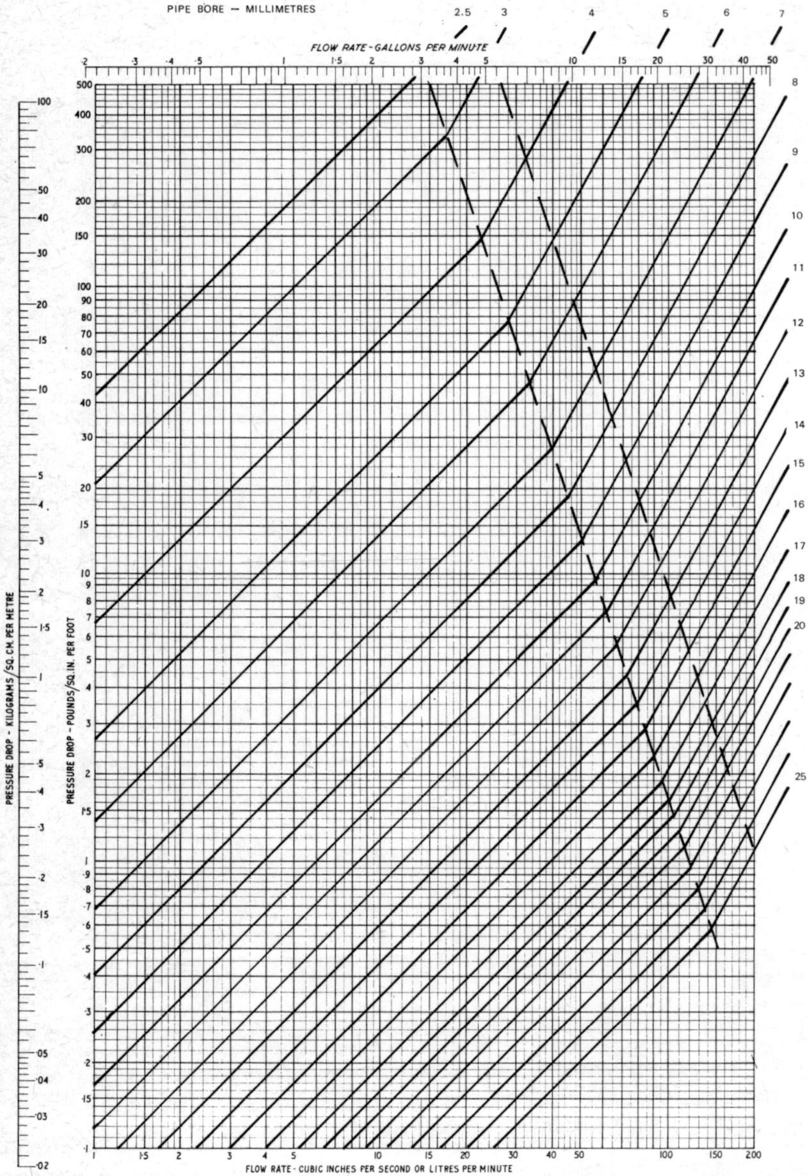

VISCOSITY 70 CENTISTOKES

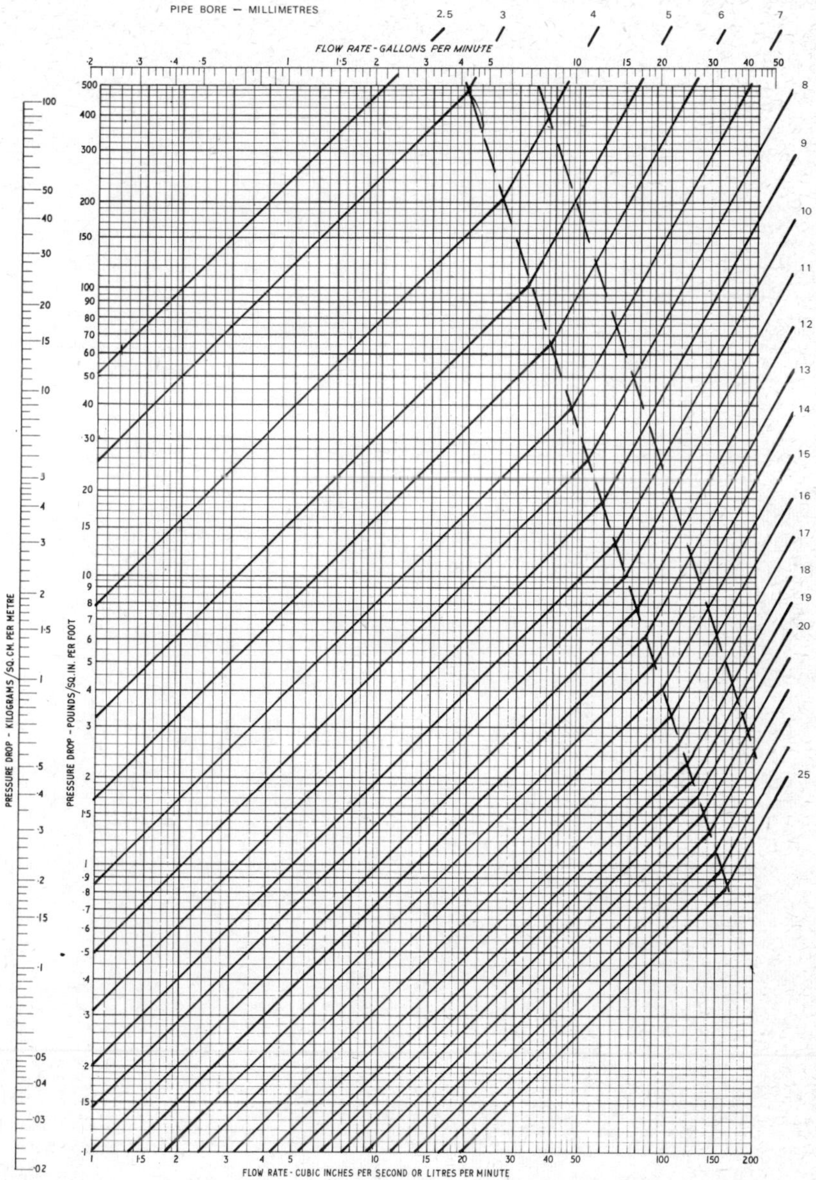

VISCOSITY 80 CENTISTOKES

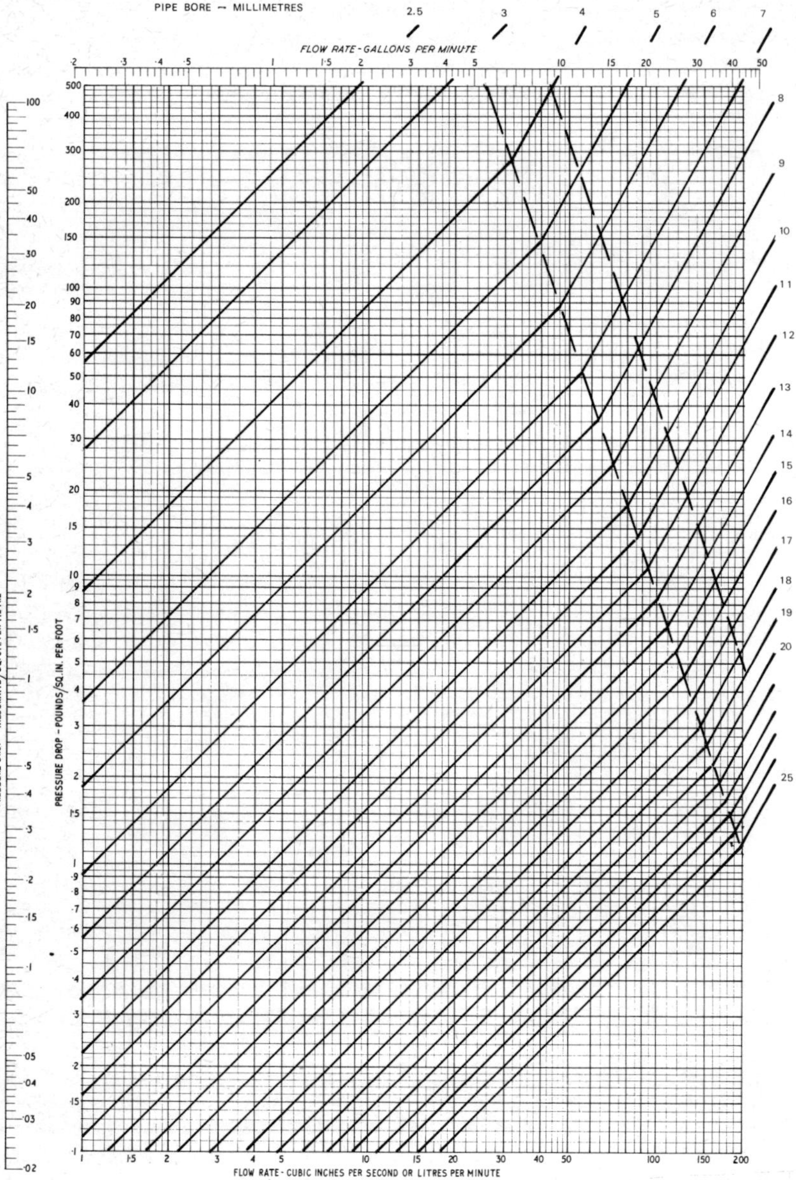

VISCOSITY 90 CENTISTOKES

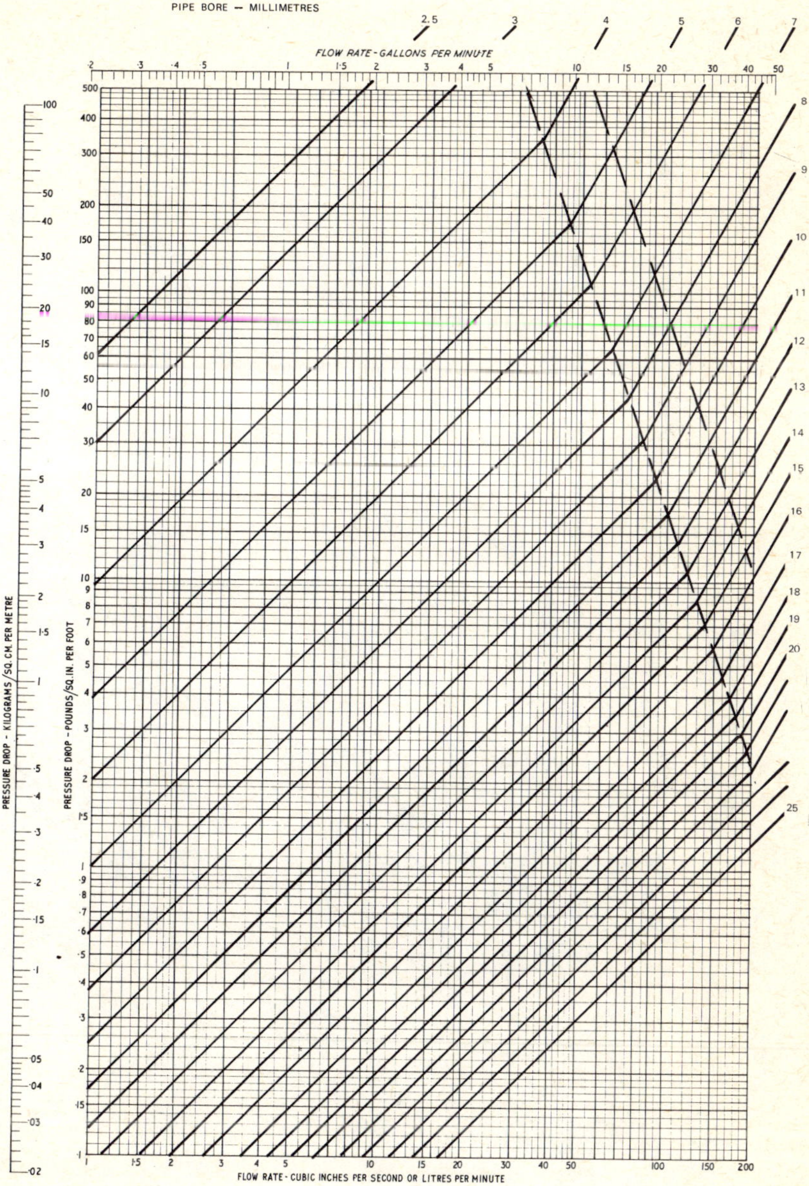

VISCOSITY 100 CENTISTOKES

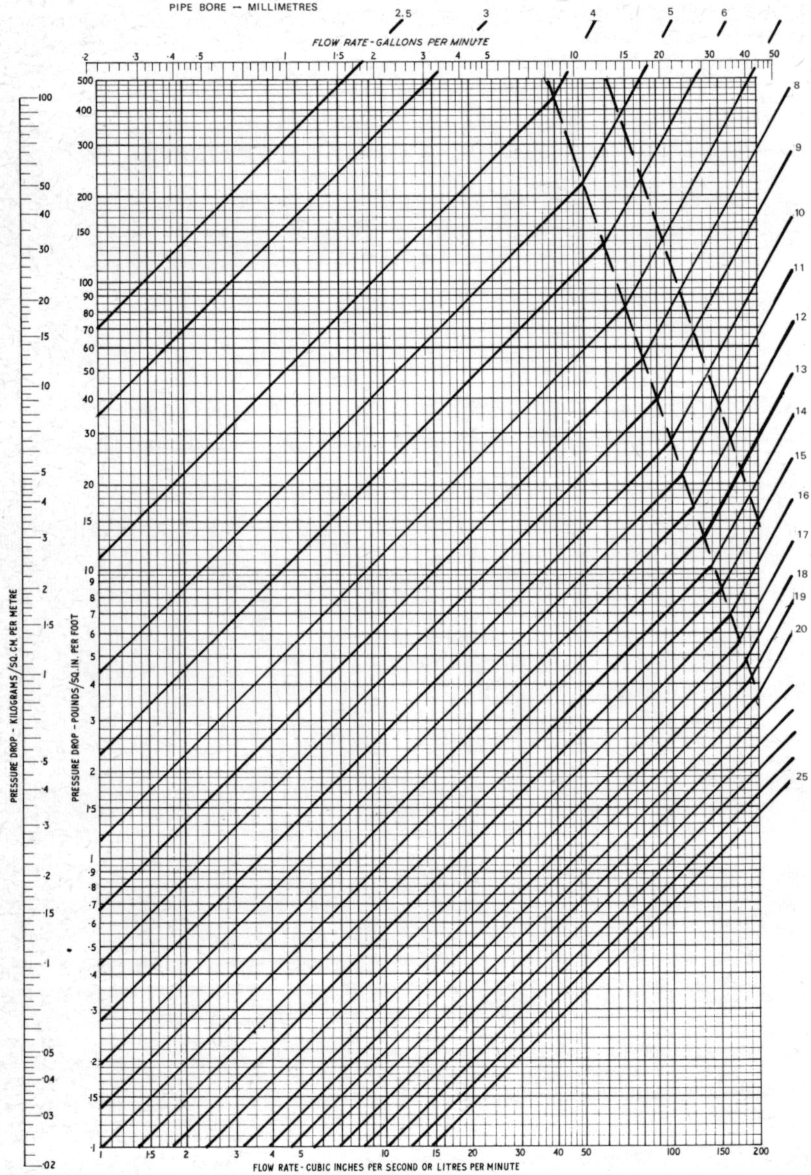

INDEX

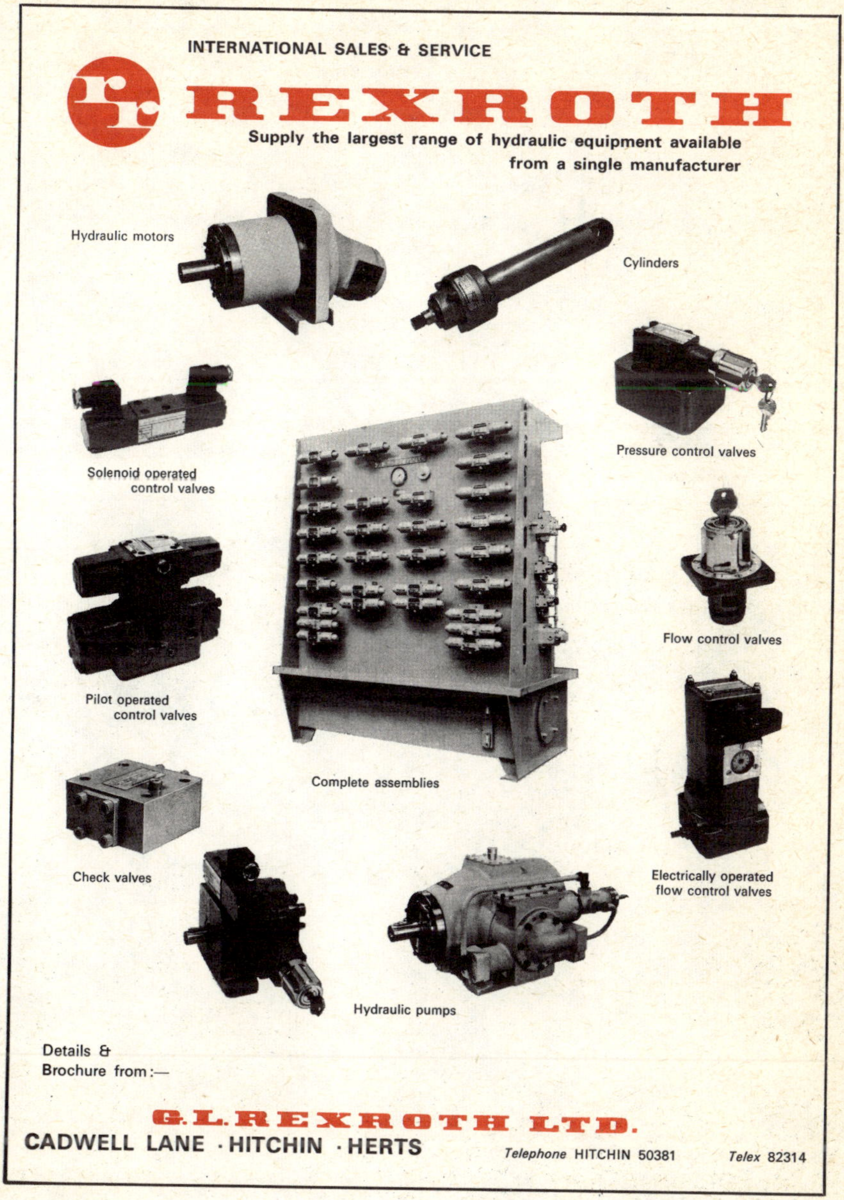
INTERNATIONAL SALES & SERVICE
REXROTH
Supply the largest range of hydraulic equipment available from a single manufacturer
Hydraulic motors
Cylinders
Pressure control valves
Solenoid operated control valves
Flow control valves
Pilot operated control valves
Complete assemblies
Check valves
Electrically operated flow control valves
Hydraulic pumps
Details & Brochure from:—
G.L.REXROTH LTD.
CADWELL LANE · HITCHIN · HERTS
Telephone HITCHIN 50381
Telex 82314

INDEX TO ADVERTISERS